WILD PLACES UK

The UK's Top 40 Nature Sites

To Ceri, Dewi and Tomos

IOLO WILLIAMS

WILD PLACES UK
The UK's Top 40 Nature Sites

SEREN

SEREN

is the book imprint of
Poetry Wales Press Ltd.
Suite 6, 4 Derwen Road,
Bridgend. CF31 1LH

www.serenbooks.com
facebook.com/SerenBooks
twitter: @SerenBooks

The right of Iolo Williams to be identified
as the author of this work has been asserted in
accordance with the Copyright,
Designs and Patents Act, 1988.

ISBN 978-1-78172-521-4

A CIP record for this title is available
from the British Library.

The publisher works with the financial
assistance of the Welsh Books Council.

Author website: www.iolowilliams.co.uk

Front cover photograph: Puffins at sunset,
Skomer, Pembrokeshire by George Wheelhouse
www.georgewheelhouse.com

Back cover photograph: Teesdale by
Thomas Thompson

Printed by Severn, Gloucester.

CONTENTS

INTRODUCTION

It isn't an easy task to choose just 40 nature sites from a list of hundreds and whichever sites I choose, I am well aware that it will spark a fierce debate around the country. I hope, therefore, that I have chosen wisely, as I have tried to include a geographical spread as well as a variety of habitats and landscapes.

Many of the reserves will be well known to most of you but I hope I have included a few surprises. There are some glaring absentees and I am well aware that certain regions are better represented than others, but this is partly because I know some parts of the UK better than others and I am ashamed to say that I have yet to visit some noteworthy sites that could, and perhaps should, be included.

You will notice that the title states 'nature sites' and not 'nature reserves', and that some of the locations in this book are rather large. This is because, for areas such as Mull for example, I find it impossible to recommend just one or two restricted areas when the whole island is a haven for nature and a visitor could stumble across incredible wildlife virtually anywhere.

If I have offended some people by omitting their favourite reserve, I apologise as that is not the purpose of this book. It has been written in an attempt to encourage people to get up off their backsides to enjoy some of the glorious wildlife we have on offer, often right on our own doorsteps. Besides, it leaves the door open for a follow-up book of 40 more wildlife sites across the country!

So much of our wildlife has disappeared since I was born more than 50 years ago and it pains me that my two sons didn't get to experience many of the natural adventures I enjoyed when I was growing up. If we can encourage more people to visit our wild areas, wherever they may be, it will engender a greater appreciation of the wildlife they hold. Once we value it, we will fight to safeguard its future. If this book helps the process, it will have achieved its aims.

It goes without saying that wherever you go, the welfare of the wildlife should always come first. Enjoy the book and, above all else, venture outside to enjoy and appreciate our incredible wildlife.

Gannets at Muckle Flugga

1. HERMANESS NNR

This spectacular reserve lies at the northern tip of Unst in the Shetland Islands and overlooks the wonderfully named Muckle Flugga, the most northerly point in Britain. Designated a National Nature Reserve in 1955, it is a haven for over 100,000 seabirds and is currently managed by Scottish Natural Heritage. Consisting of huge sea cliffs and open moorland, it encompasses the whole 965 hectares of the Hermaness peninsula and provides some of the most spectacular scenery in the UK.

The majority of visitors come to see the thousands of breeding seabirds in spring and early summer. Eroded by the might of the Atlantic Ocean, the sheer cliffs, reaching 170 metres in height, provide relatively safe nesting sites for internationally important seabird assemblages of fifteen different species.

A boardwalk leads across the moorland to the sea cliffs on the west coast and it is here that one of the reserve's most famous residents, the bonxies or great skuas, are most visible during spring and summer. Between 800-900 pairs nest on the reserve's open moorland, a far cry from the 3 pairs present in 1831 when conservation efforts to protect this species were instigated.

Great skua

The blanket bog is also home to nesting dunlin, golden plover, snipe, curlew, skylark and wheatear and the small, isolated bog pools and lochans provide the seclusion required by the more elusive red-throated divers. Cotton grass, sundew and butterwort are all common in the wetter parts of the moor whereas the more sheltered eastern side of the reserve is home to three species of rare hawkweeds, two of which are only found in Shetland. Chickweed wintergreen and northern bilberry can also be found here.

Sundew

Perhaps the most eye-catching birds at Hermaness are the more than 16,000 pairs of gannets that transform the cliffs and stacks during the summer months, the sixth largest gannetry in Britain. These large seabirds share the cliffs with breeding fulmar, guillemot, kittiwake, razorbill and shag whereas puffins breed in burrows in the soft turf along the cliff tops. Although gannet numbers are increasing on the reserve, the populations of most other seabirds are in decline, almost certainly due to changing sea temperatures and the overfishing of sandeels, a favoured prey.

Ravens and rock doves also nest along the tall cliffs and seasonal colour is provided by salt tolerant plants such as red campion, thrift, Scots lovage and wild angelica. 53 species of beetle and 46 different spiders have been recorded as has the rare northern arches moth. Offshore, sightings of harbour porpoise, minke whale and orca are relatively frequent and both grey and common seals are resident along the coastline.

To reach Hermaness, follow the B9086 from Haroldswick and follow the road as far as it goes. A final left fork takes you to the car park, the right fork to the visitor centre in a converted lighthouse. The walk from the car park to the seabird cliffs is not easy but please keep to the paths to avoid disturbing breeding birds and keep away from the cliff edges. Victoria's Vintage Tea Rooms in nearby Haroldswick does great food and fantastic cakes.

The Hermaness Gannetry

2. FETLAR

Aptly named the garden of Shetland, Fetlar is a relatively fertile island of abundant wildlife and stunning scenery. Situated in the north east of the archipelago, it is Shetland's fourth largest island at just over 4,000 hectares in size, but it has a population of just 61.

Unlike the other islands, large parts of Fetlar are covered in wet pasture as well as moorland, making it one of the best places in the UK to see breeding waders. Indeed, visitors are often greeted at the ferry terminal by drumming snipe, a summer sound that accompanies you throughout the island. Redshank and lapwing also breed here in good numbers, as do dunlin, golden plover and curlew.

Fetlar's star bird, however, is the dainty red-necked phalarope with more than half the UK's population found here. The Loch of Funzie and the nearby Mires of Funzie at the east end of the island are the best places to see these birds although they can prove to be remarkably elusive. It's well worth spending time in the hide at the Mires of Funzie, not only to look for phalaropes, but also teal, snipe and redshank as well as a host of plants including marsh cinquefoil and tussock sedge.

Red-necked phalarope

The area around the car park at the Loch of Funzie is also good for arctic skuas and whimbrel, both of which breed in small numbers on the nearby moorland. Red-throated divers, arctic terns and great skuas are frequent visitors to the loch and in summer, the surrounding fields are full of heath spotted orchids.

The airstrip road at Vord Hill is a good place to see a selection of breeding waders including golden plover, dunlin, ringed plover, curlew and whimbrel, and arctic skuas nest here in small numbers. Numbers of the latter two species are declining on the island therefore it's very important that visitors stay on the track and don't wander out onto the moor. It was on Stakkaberg at Vord Hill that the famous Shetland naturalist, Bobby Tulloch, found a breeding pair of snowy owls in 1967. Unfortunately, the owls are long gone.

Dunlin

Papil Water, a coastal loch behind Tresta Beach, is worth visiting for its large number of bathing great skuas and to witness arctic skuas harassing arctic terns as they bring food back to the nest. Eider ducks can often be seen here with their chicks and the shore is good for visiting sanderling, turnstone and dunlin. Fetlar's rarest plant, water sedge, can be found along the margins of Papil Water and autumn gentian grow along Tresta links. The beach at Tresta is a great place to go in search of quahog shells, a clam that lives in deep waters in the north Atlantic.

An invasive plant, the monkeyflower, adds colour to many of Fetlar's wet ditches in late summer and the road verges are often full of creeping buttercup, red campion, meadowsweet and birds' foot trefoil. These flowers attract the beautiful Shetland bumble bee, a large orange and yellow bee that is confined to these islands and the Outer Hebrides.

Common and grey seals can both be seen around the coast and it's worth keeping a constant eye open for otters. Offshore, pods of orcas or a passing minke whale or risso's dolphin are also possible. Gannets, black guillemots, puffin, fulmar and eider ducks are all commonly seen out at sea, often on the ferry crossing from either Gutcher on Yell or Belmont on Unst. Storm petrel also breed around the coast whilst the island was formerly the Shetland stronghold for breeding Manx shearwater.

The small village shop at Houbie has a small café that is open all summer and sells wonderful home made cakes.

Common seal

3. FORSINARD FLOWS RSPB RESERVE

Situated in Caithness and Sutherland in the far north east of mainland Scotland, Forsinard Flows is part of a vast expanse of blanket bog, sheltered straths and mountains known as the Flow Country. The largest expanse of blanket bog in Europe, this mosaic of peat bog, pools and lochs is home to a rich variety of wildlife, much of which can be seen on this 21,000ha RSPB reserve.

This unique landscape is a mosaic of land and water and the pools and lochs provide perfect breeding and feeding grounds for black-throated and red-throated divers although sadly, these two species are not visible from the walkway or the excellent viewing tower looking out over the reserve. Red-throated divers may be seen on the Forsinain trail and on lochs to the south of the site whereas black-throated divers are confined to the larger lochs to the west of Forsinard.

In spring, visitors are far more likely to see and hear golden plover and dunlin, both being widespread breeders throughout the extensive areas of blanket bog. A far scarcer breeding wader, the greenshank, can also be seen on and around the reserve as can hen harrier, merlin, short-eared owl, and skylark. Dipper and common sandpiper feed along the rivers, streams and lochs and the occasional raven, peregrine falcon and golden eagle can sometimes be seen soaring above in search of prey. Wild greylag geese, teal and wigeon breed on and around some of the lochs and lochans.

Hen harrier

The rare common scoter nests in diminishing numbers, as does the occasional wood sandpiper, and the elusive otter is present, although rarely seen. Red deer herds are particularly visible in this open landscape and in autumn, they can be seen and heard rutting from the Dubh Lochan trail. On sunny summer days, the boardwalk is a good place to look for basking common lizards and dragonflies aplenty, including large red and common blue damselflies and the azure hawker. The large heath butterfly can also be spotted flying low over the moorland on still, warm days in early summer.

Large red damselfly

Common blue damselfly

Mountain hares and foxes are present and the hundreds of lochs and fast-flowing streams and rivers on and around the reserve are home to brown trout, sea trout, Atlantic salmon and Arctic charr as well as important populations of the freshwater pearl mussel. The most important plant species are the sphagnum mosses that bring life to the active peat bogs and provide suitable habitat for cotton grass, bearberry, dwarf birch, bog orchid, bog asphodel and carnivorous plants such as butterwort and sundew.

The Forsinard Flows reserve is situated alongside the A897 road in Strath Halladale and the RSPB visitor centre is located in the railway station building on the Far North line between Inverness to the south and Thurso and Wick to the north.

4. ANAGACH WOODS

Despite being located only a stone's throw away from its larger and more celebrated cousins, Abernethy and Rothiemurchus, Anagach does not suffer an inferiority complex, and rightly so. This delightful, 386ha open Scots pine woodland is located between the town of Grantown-on-Spey and the River Spey in the shadow of the Cairngorm Mountains and dates back to 1766 when Grantown was established as a new Highland industrial town.

The woods support a very healthy population of red squirrels, some of which are attracted to nearby gardens and to feeders erected along one of the trails. Roe deer can frequently be seen grazing the woodland floor vegetation, particularly at dawn and dusk, and although pine marten are present, their scats are often the only evidence of them.

Red squirrel

Otters can sometimes be seen on the River Spey on the eastern edge of the woodland, along with birds such as dipper, grey wagtail, common sandpiper, goosander and goldeneye, all of which breed here. The river is also a good place to keep an eye open for passing ospreys in spring and summer as they make their way to and from nearby nesting sites.

With so many mature Scots pines in the woodland, it is an excellent place to see some of the Scottish pinewood specialities such as common and Scottish crossbills and crested tit, the latter often sharing the feeders with the red squirrels. Treecreepers, siskin and redstart are also regularly recorded and although capercaillie are seen on occasion, they are now extremely scarce and should, wherever possible, be left in peace.

Male common crossbill

Crested tit

Much of the woodland floor is covered in heather and bilberry but botanical gems such as the delicate twinflower can be found in some places and this is the largest inland colony of the 6-spot burnet moth in the Highlands. Nests of hairy wood ants can be seen here, as can the scarcer but wonderfully named slave-maker ant.

Twinflower

Anagach is a community woodland and is well-walked towards the south-western end nearest the town but with so many waymarked trails of varying lengths, wildlife lovers can easily escape to the quieter areas. There are a few good cafés in Grantown itself but the best is the 'High Street Merchants'. The wood is easily reached from the centre of Grantown-on-Spey.

5. ABERNETHY NATIONAL NATURE RESERVE

No other reserve in Britain conjures up images of wildwood full of prowling wolves, bears and lynx quite like Abernethy. Unfortunately, the large predators have long gone but this 12,754ha National Nature Reserve still holds the largest remaining single area of native Caledonian pine woodland in the UK.

Situated in Strathspey and within the boundary of the Cairngorms National Park, the vast majority of the reserve is owned and managed by the RSPB with a small section, Dell Wood, managed by Scottish Natural Heritage (SNH). Stretching from the River Nethy to the top of Ben Macdui, the NNR encompasses a variety of habitats including open moorland, fast-flowing rivers and streams, mountain plateaus, lochs, fens and bogs as well as Caledonian pinewood. With so many paths and trails, visitors of all abilities can access most parts of the reserve without disturbing the wildlife, an important consideration for a place containing so many rare species.

Over 200 species of vascular plants have been recorded here including several nationally scarce species such as twinflower, intermediate wintergreen, creeping lady's tresses and heath cudweed. The assemblages of lichens and fungi are exceptional and include stump lichen and 11 species of tooth fungi. In addition, Abernethy is amongst the most important invertebrate sites in Scotland with species such as narrow-headed ant, pine hoverfly, pearl-bordered fritillary and the cousin German moth as well as northern damselfly, white-faced darter and northern emerald.

With much of the reserve under RSPB ownership, Abernethy is rightly famed for its birds. It is home to several ancient pinewood specialists such as capercaillie, crested tits and Scottish crossbills although all these birds can prove to be remarkably elusive, particularly the capercaillie which is now in danger of extinction across much of its former range in Scotland. Other woodland birds include black grouse, redstart, spotted flycatcher and tree pipit as well as common crossbill and parrot crossbill.

The rivers and streams are home to dippers, grey wagtails, common sandpiper and goosander and ospreys can be viewed from the famous Loch Garten osprey centre. Golden and white-tailed eagles are occasionally seen, the former particularly towards the higher moors and mountains, as are peregrine falcons and goshawks.

Wildcat

Osprey

Pine marten

Red and roe deer are common and most visible at dawn and dusk whereas red squirrels are active throughout the day and best seen at the bird feeders near the osprey centre. Otters are occasionally spotted along the River Spey and although pine marten are widespread, they are elusive and strictly nocturnal. The holy grail for any mammal enthusiast, however, is the wild cat and although small numbers are known to be present, like the capercaillie, they are rarely seen.

The Osprey Centre at Loch Garten is open between April-August and has a few drinks and a well-stocked shop, but places to eat can be found at the nearby villages of Nethy Bridge and Boat of Garten.

Capercaillie

6. INSH MARSHES

One of the most important wetlands in Europe, Insh Marshes is located between Kingussie and Kincraig in Badenoch and Strathspey. With the Cairngorm mountains to the east and the Monadhliath hills to the west, this 10 square kilometres of River Spey floodplain provides a refuge for an incredible variety of wildlife.

Partly due to seasonal flooding, the reserve consists of swamp, fen and carr as well as deciduous woodland, grassland and scrub. Adding to the diversity of habitats, the reserve also includes parts of the River Spey and Loch Insh and is recognized as the best example of an unspoiled river floodplain in the UK.

Sedges, reed-bed, herb-rich swamp and willow-carr wetland dominate a site that is the UK stronghold for water sedge and string sedge. Other rare plants include the least yellow water lily, awlwort, cowbane and the shady horsetail. It is the best site in Scotland for rare water invertebrates including several species of flies, an endemic reed beetle and a wetland spider *Wabasso replicatus* found nowhere else in the UK, and it boasts an incredible 51 species of caddis fly. The aspen hoverfly and two rare moths, Rannoch sprawler and cousin German can all be found in some of the broadleaved woodlands that fringe the marshes and this is one of only 4 sites in the UK for the dark bordered beauty moth.

Cousin german

Loch Insh supports a breeding population of Arctic charr that is unlike any other Scottish form of charr and closely resembles the fish found in Lake Windermere. The area is also important for its population of otter and the Lookout next to the RSPB car park is a good place to look for them. Roe deer graze out on the open marsh and red squirrels are common in the scattered woodlands around the periphery.

During April to June, the wetlands are alive with displaying lapwing, redshank, curlew and snipe, and good numbers of wigeon and wild greylag geese also breed on the marshes. Sedge warbler, grasshopper warbler and water rail are common but spotted crakes have become increasingly irregular in recent years.

Marsh harriers are seen here in summer and have bred in the past, and there is a well-known and well-photographed osprey nest at the edge of Loch Insh on the outskirts of Kincraig. Goosander, dipper, grey wagtail and common sandpiper all nest along the River Spey and this area supports a significant percentage of the UK population of breeding goldeneye.

Winter brings whooper swans as well as pink-footed geese, teal and wigeon, and this is the best season to look for raptors on the reserve. Peregrines and merlin both hunt birds over the marshes and hen harriers gather to hunt small mammals and to roost in the taller vegetation. White-tailed eagles are becoming increasingly common and diligent scanning over the adjacent mountains could reward the observer with a distant view of a golden eagle.

The impressive hide alongside the B970 near the Ruthven Barracks gives excellent views over the wetlands and has disabled access. The nearby town of Kingussie has cafes and other facilities.

Goldeneye

Redshank

7. CAIRNGORMS

The Cairngorms mountain range is a vast upland area in the eastern Highlands that became Scotland's second National Park in 2003. It consists of high plateaux at between 1000-1200 metres above sea level with domed summits rising above them to a height of around 1300 m.

The mountain range forms an arctic alpine mountain environment with tundra-like characteristics and long-lasting snow patches. Much of the landscape is boulder-strewn and steep cliffs of granite are found around the edges in many areas. This is a unique landscape in UK terms and as such, it supports very unusual communities of flora and fauna.

These mountains are home to extensive tracts of arctic-alpine plant communities with excellent lichen-rich heaths, snow bed communities and montane bogs. Rare plants include alpine sow thistle, trailing azalea and alpine milk-vetch as well as woolly willow and woodsia, a genus of ferns commonly known as cliff ferns. Several nationally scarce mosses and lichens can also be seen here, including snow fork-moss, a handsome moss that forms yellow-green cushions, often amongst boulders. As well as being important for a range of cold-loving invertebrates, the Cairngorms are home to Britain's only truly montane species, such as the Arctic whorl snail and the Scottish mountain spider.

Alpine sow thistle

Alpine milk vetch

Ptarmigan

Red grouse

Red deer can be found on some of the lower, heather-clad slopes whereas Britain's only herd of reindeer roam the high mountains having been re-introduced by a Swedish herdsman in 1952. The only truly wild montane mammal found here, however, is the mountain hare which changes colour from various shades of brown in summer to almost pure white in winter, a perfect adaptation for such a snowy environment.

Most naturalists walk the hard yards up onto the high plateaux in search of the unique breeding bird community. Ptarmigan, snow bunting and dotterel all nest on the high tops whereas golden eagle and peregrine falcon patrol the skies overhead. Ring ouzels breed in many of the rocky gullies and red grouse can be seen and heard on some of the lower slopes.

The high tops of the Cairngorms are not a place for the fainthearted and even in summer, this can be a dangerous habitat. Visitors must be aware that access is not possible from the top of the funicular railway, therefore reaching the high tops requires a great deal of effort and suitable equipment. The Ptarmigan café at the Top Station of the funicular railway serves food and drinks otherwise there are plenty of places to eat in the surrounding towns and villages.

8. BALRANALD

The Balranald nature reserve is an RSPB reserve situated on the north-west coast of North Uist in the Outer Hebrides. Sculpted by a combination of Atlantic storms and generations of traditional crofters, the reserve consists of golden beaches, rocky shores, sand dunes, flower-rich grasslands and wetlands.

Without a doubt, Balranald's star bird is the corncrake, a species that has suffered a catastrophic decline over much of its former European range. Here, it has survived and thrived in the flower-rich hay meadows that are harvested in late summer. The birds are incredibly elusive, however and are easiest to see in late April and early May, soon after returning from Africa and before the vegetation grows too tall.

Corncrake

In spring and early summer, the flower-rich machair on the landward side of the dunes and the wet pastures are alive with the sound of displaying redshank, lapwing, dunlin and snipe, all of which breed in exceptionally high densities here. Ringed plover and oystercatcher also nest on the reserve, as do corn bunting, and arctic tern nest along the rocky shore.

In autumn and winter, flocks of turnstone, purple sandpiper and sanderling gather along the shoreline whereas barnacle geese join large numbers of lapwing and golden plover on the machair. Mixed flocks of skylark, twite and snow bunting attract hunting merlin and peregrine and the occasional golden and white-tailed eagle pass through on the lookout for prey.

The flower-rich machair is dominated by red clover and silverweed in summer although poppies, corn marigolds, early marsh orchid and Hebridean spotted orchids also catch the eye. This plethora of flowers attracts rare pollinators such as the northern collettes bee, the great yellow bumblebee and the moss carder bee as well as the nationally rare belted beauty moth.

Moss carder bee & Scotch argus

Otters and grey seals are frequently seen along the coastline and great skuas harass the arctic terns as they return to their colonies with fish. Gannets and red-throated divers can also be observed out at sea and in spring and autumn, the shoreline is a great place to watch seabird passage.

Great yellow bumblebee

Balranald is found 3 miles north of Bayhead and is signposted off the A865. The reserve and the small visitor centre are open all year round.

White tailed eagle

Fulmar

Leach's petrel

9. ST KILDA

St Kilda would certainly win the prize for the most isolated site included in this book, as it is situated 64 kilometres northwest of North Uist out in the Atlantic Ocean. Owned by the National Trust for Scotland, the archipelago consists of 4 main islands (Hirta, Dun, Boreray and Soay), and although there has been no permanent population on St Kilda since 1930, the islands still accommodate military personnel, conservationists and volunteers.

Designated a World Heritage site in 1986, it is one of the few in the world to hold joint status for both its natural and cultural qualities, sharing this honour with sites such as Machu Picchu in Peru. Visiting yachts can find shelter in Hirta's Village Bay but anyone wishing to land has to contact the NTS in advance due to the concerns about introducing non-native plants and animals to such a fragile environment. Once the preserve of hardy adventurers, visiting is now easier as fast boats go out for the day from the Western Isles and Skye when it's calm enough.

The world's second largest gannetry numbering some 60,000 pairs is found here, albeit on three different islands within the archipelago, Boreray, Stac Li and Stac an Armin. St Kilda also holds 90% of the European population of Leach's petrels, around 49,000 pairs, as well as an incredible 136,000 pairs of puffins. When you add 65,000 pairs of fulmar to this total, it's no wonder that seabirds formed an important part of the inhabitants' diet for several centuries.

Guillemot and razorbill also breed here in good numbers, making the most of the tallest sea cliffs in Britain, and more than 150 pairs of great skua nest on the flatter turf-covered tops, particularly on Hirta. The last great auk seen in Britain was killed on Stac an Armin in July 1840, a mere four years before the last ever breeding pair was killed off Iceland.

The St Kilda wren, a subspecies of the Eurasian wren, is unique to these islands, as is the St Kilda field mouse, a subspecies of the wood mouse. A third unique subspecies, the St Kilda house mouse, died out when the last inhabitants abandoned the islands in 1930.

Due to its isolation, there is little diversity in the flora and fauna of the island although several migrant invertebrates such as the silver-Y moth and red admiral butterfly are recorded most years. The St Kilda dandelion is an endemic species first identified in 2012 and one endangered weevil, *Ceutorhynchus insularis*, is known only from Dun and the Westermann Islands off the south-west coast of Iceland.

The marine life around the archipelago attracts divers from all over the world to witness a kaleidoscope of form and colour as sea anemones, soft corals, sponges, hydroids and bryozoans all thrive in the plankton-rich waters. Grey seals patrol the inshore waters as does the occasional basking shark and minke whale.

Due to the unpredictable weather, there is no guarantee that visitors can get across to St Kilda but the effort, if you make it, is well worthwhile.

St Kilda wren

10. ISLE OF MULL

With an area of 875 square kilometres, Mull is Britain's fourth largest island and lies off the west coast of Scotland in the Inner Hebrides. Most of the small resident population of fewer than 3,000 people live in the town of Tobermory on the east coast, leaving vast tracts of the countryside to the wildlife.

A combination of 480 kilometres of coastline, mountains reaching almost 1,000 metres above sea level, vast tracts of moorland, substantial woodlands and wet pastures means that unexpected wildlife encounters can occur anywhere, hence its inclusion in this book. Indeed, the island has earned itself an enviable reputation as Britain's premier wildlife tourism destination and during the summer, Mull's population can more than triple due to the numbers of visitors.

The waters around Mull are rich in wildlife and in summer, minke whales, basking sharks, harbour porpoise and bottle-nosed dolphins are regularly seen, particularly around the northern and western headlands. Both grey and common seals occur along the coastline and the shorelines are some of the best places south of Orkney to see otters. Salen Bay and the eastern end of Loch na Keal are good for otter and seals.

Otter

Mull is often known as eagle island, and for good reason. More than 30 pairs of golden eagles and over 20 pairs of white-tailed sea eagles nest here, making it one of the best places in the UK to see these large raptors. Both species can be encountered virtually anywhere but a Mull Eagle Watch scheme is set up each year to view a nearby sea eagle nest. Golden eagles are more elusive despite being more numerous but the peaks around Ben More are worth scrutinizing with binoculars.

Golden eagle

Hen harriers also flourish on Mull thanks to a lack of persecution and birds can be seen throughout the year over areas of moorland and rough pastures such as beside the road to Lochbuie or south-east of Calgary. Short-eared owls often frequent the same habitat as the harriers but their numbers fluctuate widely in response to the field vole population. Kestrels and buzzards are common and sparrowhawks can often be seen near the large conifer plantations. Peregrines, however, are surprisingly scarce.

Seabirds such as gannets, puffins, razorbills, guillemots and black guillemots are plentiful offshore, on occasion accompanied by a predatory great skua. Small numbers of red-throated divers breed on isolated lochs and in winter, internationally important numbers of great-northern divers and smaller numbers of black-throated divers and slavonian grebes gather around the coast and on many of the sea lochs. Eider duck, red-breasted merganser, shag and cormorant all breed around the coast.

Whinchat, wheatear and twite breed on the island as do lapwing, snipe and golden plover. Although corncrakes are now only found sporadically on Mull, they do nest on the island of Iona, a short ferry ride away.

Mull has over 800 species of vascular plants including 33 species of fern and at least 18 orchids, including the narrow-leaved helleborine. There are around 700 species of lichen, including the scarce tree lungwort, and more than 2,000 species of fungi, including the rather peculiar hazel glove fungus.

There are several ways of reaching Mull but the most popular is to take the ferry from Oban to Craignure. It does get very busy during the summer months and it is advised to book well in advance. There are plenty of good general stores and gift shops around Mull as well as numerous coffee shops and restaurants with Tobermory having the widest choice.

Hazel glove fungus

11. LOCH GRUINART

The most striking geographical feature on the sparsely populated north coast of Islay in the southern Hebrides is Loch Gruinart, a shallow sea loch that extends 6 kilometres inland. Sheltered from the worst of the Atlantic storms by the promontory at Ardnave point, the site has long been a refuge for people and wildlife.

Located around the head of the loch is the 1667 hectare RSPB reserve that encompasses saltmarsh, mudflats, wet grassland, woodland and moorland. Such a variety of habitats attract a rich diversity of wildlife throughout the seasons, although Loch Gruinart is probably best known for its wintering birds.

In autumn, up to 35,000 Greenland barnacle geese (45% of the global population) and around 1,000 Greenland white-fronted geese arrive to spend the winter months here, as do large numbers of pintail, shoveler, golden plover and black-tailed godwit. The two reserve hides give excellent views of these birds as well as raptors such as merlin, hen harrier and peregrine falcon that are attracted by the abundance of prey.

Lapwing

As a result of the dedicated in-house farming operation, in spring and summer, the lowland fields can be managed for breeding waders. More than 200 pairs of lapwing and over 100 pairs of redshank nest here, as do several pairs of curlew and snipe. The provision of plenty of early cover means that corncrake do well on the reserve and their monotonous, rasping calls can be heard on calm evenings from early May onwards. This is also a good time to look out for boxing hares as females repel the advances of amorous males.

Speckled wood butterfly

Spring is the best time to take a walk through the broadleaved woodland as resident birds such as treecreepers and great tits are joined by summer migrants like the willow warbler. In May, the woodland floor is a carpet of bluebells with the occasional red campion and late-flowering primrose scattered amongst them. Speckled wood butterflies are also found here as are roe deer, although these shy animals will venture out onto adjacent fields in the evening.

The moorland trail takes you out onto higher ground, the haunt of breeding hen harriers and short-eared owls and the occasional passing golden eagle. Carnivorous plants such as sundew and butterwort thrive in these acidic, peaty conditions and an abundance of devil's bit scabious supports a thriving colony of marsh fritillary butterflies.

The focal point for the reserve is the impressive visitor centre that is open all year. This is to be found as part of a complex of farm buildings just south west of the loch itself. The car park for the hides and various trails is located along the narrow road running up the loch's western side towards Ardmore.

Peregrine falcon

12. BASS ROCK

The Bass Rock is a small island, approximately 2 kilometres offshore, in the outer part of the Firth of Forth in the east of Scotland. It is a steep-sided volcanic rock, 120 metres at its highest point and was, historically, home for an early Christian hermit, a prison and the site of an important castle.

The 3 hectare island is now uninhabited but it plays host to the world's largest gannetry. 150,000 gannets make their home here during the spring and summer months, producing an incredible 152,000kg of ammonia every year! In late summer, when the adults are feeding their growing chick, the island is a hive of activity and it's no wonder that it was recently described by Sir David Attenborough as one of the 12 wildlife wonders of the world.

Nesting gannets

As gannet numbers increased and spread from the cliffs onto the flatter top of the island, they displaced the herring and lesser black-backed gulls and have now run out of space. As a consequence, gannets have recently started breeding at Bempton, Troup Head and St Abbs.

Although the gannets are undoubtedly the main attraction, other bird species that frequent the island include guillemot, razorbill and eider duck. Kittiwake, fulmar and shag also nest on this ancient volcanic plug and peregrines are often seen patrolling the skies above. Puffins formerly nested in good numbers but are now confined to several nearby islands.

Razorbill

Grey seals fish around the base of the rock and several species of whales including minke and humpback have been observed in the surrounding waters. Bottlenose dolphins and harbour porpoises are regularly seen here but orcas are less frequent although several have been seen in recent years.

The soil is fertile and supports a wide variety of vascular plants such as thrift and sea campion. Tree mallow, a flowering plant that thrives on guano, is also found here.

Boat trips around and landing on Bass Rock can be arranged through the Scottish Seabird Centre at North Berwick. There is, however, no guarantee of landing due to changing weather conditions, and there are no toilet facilities on the island. Several cafés and restaurants can be found in North Berwick.

Fulmar

13. WOOD OF CREE

Located north west of the town of Newton Stewart in Dumfries and Galloway, Wood of Cree is the largest ancient woodland in southern Scotland. Owned and managed by the RSPB, recent additions to the reserve and the planting of more than 200,000 new trees mean that visitors will soon be able to walk through 18 kilometres of uninterrupted deciduous woodland.

The large Atlantic oak woodland, also known as a Celtic rainforest, is the main feature of the reserve but it also includes other important habitats such as wood pasture, wet floodplain woodland, open moorland, fen, scrub and grassland. As one would expect, this variety of habitats attracts an impressive array of flora and fauna.

Spring is the best time to visit the reserve as the woodland floor fills with a carpet of bluebells and migrant birds arrive back from Africa. Indeed, it is said to be one of the best bluebell woods in the whole of Scotland, although there are plenty of other notable plant species here also. Scarce species such as whorled caraway and globeflower can be seen in spring, although the latter is now very rare, and Wilson's filmy fern thrives in the wet conditions. The wood has a great selection of sedges including the rare ginger-bread sedge and the scarce water sedge.

Globeflower

Ginger-bread sedge

Red squirrels are regularly seen as are roe deer and although otters and pine marten are present, they are both elusive. Eight species of bat are found on the reserve, including Daubenton's and the rare Leisler's bat, the perfect combination of a plentiful supply of invertebrates and roosting sites proving vital for their success.

Purple hairstreak butterflies flit between the leaves at the top of the oak trees in summer and the Scotch argus is on the wing around the same time, particularly along the scrubland trail and in the wood pasture. The delicately patterned small pearl-bordered fritillary can also be seen on warm days in early summer.

Scotch argus

Purple hairstreak

It is the birds that attract most visitors to the site however, particularly in spring. Resident species such as treecreeper and willow tits are joined by pied flycatchers, redstarts, wood warblers and tree pipits. The distinctive reeling song of the grasshopper warblers can be heard in the wood pasture in summer as can the call of the cuckoo and singing whinchats. Black grouse are often recorded around the woodland edges, particularly at dawn and dusk.

Barn and tawny owls are both resident and once the summer migrants have departed for warmer climes, they are replaced by wintering species such as teal, mallard, goldeneye and whooper swans. Winter is a good time to see some of the woodland birds here as the leaves have gone and mixed flocks of tits are often joined by other species such as goldcrest and woodpeckers.

Barn owl

To reach the reserve, travel north along the minor road from Newton Stewart through Old Minnigaff. Turn left past the church and continue along the minor road for a further 5 kilometres to the car park. There are no facilities on site but Newton Stewart has a few cafes.

14. LEIGHTON MOSS RSPB RESERVE

Situated near Silverdale on the edge of Morecambe Bay in Lancashire, Leighton Moss has been an RSPB reserve since 1964 and hosts a wide variety of spectacular wildlife. It is home to the largest area of reed bed in north-west England and within its boundaries, it supports diverse habitats including woodland, limestone grassland, mudflats, coastal marsh and saltwater lagoons.

Its importance as a wildlife site is reflected in its designation as a Ramsar Site, a Site of Special Scientific Interest, a Special Protection Area and an Important Bird Area. Additionally, it sits within the Arnside and Silverdale Area of Outstanding Natural Beauty and it has been one of my favourite British reserves since my first visit in 1974.

The extensive reed beds and reed-fringed wetlands provide a home for important breeding populations of bitterns, marsh harriers and bearded tits. This reserve has long been a vital northern outpost for all three species as well as reed and sedge warblers, water rails and several wintering wildfowl and waders such as teal and common snipe.

Bearded tit

Marsh harrier

The deeper meres hold good populations of frogs, toads, newts and fish such as pike, tench, rudd and eel, and as a result, Leighton Moss is one of the best places in England to see otters in the wild. The sky tower, situated near the main entrance to the reserve, gives spectacular views over the reeds and open water and is the best place to look for these normally elusive mammals.

Bittern

Red deer are commonly seen in the reed beds and roe deer can be seen around the edges of the moss in the woodland and fen habitat. In winter, spectacular starling murmurations often attract hunting sparrowhawks and peregrines and marsh harriers are now present year-round. Growing numbers of little egrets gather in some of the wetland trees at dusk to roost and hunting barn owls can often be seen over the reed edges and rough pastures as the light fades in the evening.

Marsh tits, bullfinches, treecreepers and long-tailed tits share the mixed woodlands with colourful fungi such as scarlet elf cup, chicken of the woods, fly agaric and candle snuff. The reserve's butterflies include high brown, pearl-bordered and small pearl-bordered fritillaries, northern brown argus, dingy skipper and small heath, and almost 600 species of moths have been recorded.

Chicken of the woods

In spring and summer, the hides overlooking the Allen and Eric Morecambe pools are great places to see nesting avocets and summer-plumage black-tailed godwits stop off on the way to their breeding grounds in Iceland. Outside the breeding season, flocks of wigeon graze the saltmarsh and in autumn, substantial flocks of redshank and lapwing are often joined by smaller numbers of greenshank, knot, dunlin and spotted redshank.

Leighton Moss has a visitor centre, a well-stocked shop and a great café that sells lovely food and exquisite cakes. There are plenty of trails and hides around the reserve and it is easily reached from Junction 35 of the M6 or by rail to Silverdale station, some 250 metres from the main entrance.

15. FARNE ISLANDS

Owned by the National Trust and situated off the coast of Northumberland in north east England, the Farne Islands are a magnet to thousands of visitors every year. Consisting of between 15-20 islands, depending on the state of the tide, they lie between 2.2 and 7.6 kilometres off Bamburgh although boat trips to visit the islands depart from the nearby town of Seahouses.

In the seventh century, the Farnes were home to St Cuthbert who, legend has it, in 676, introduced special laws to protect eider ducks and other seabirds nesting on the islands. To this day, the local eiders are known as 'St Cuddy's duck'. In fact, this, in modern parlance, is fake news. It was dreamt up by a monk some 600 years after Cuthbert's death to drum up support for the legend of Cuthbert after his burial in Durham.

Puffins are, without doubt, the Farne Islands' star attractions with more than 35,000 nesting pairs. There has, however, been a recent decline in the population, in line with worrying declines elsewhere in northern Britain. 48,000 guillemots nest on the stacks and steep cliffs as do more than four thousand kittiwake and several hundred razorbill, fulmar, shag and cormorant.

Guillemots

Visitors to the Inner Farne islands are reminded to wear a hat as more than a thousand pairs of Arctic tern breed here and the many pairs that nest close to the footpaths defend their eggs and young with great vigour. Sandwich and common terns also breed as do several hundred pairs of black-headed gull and around 500 pairs of eider.

Shags

The islands hold a notable colony of around 6,000 Atlantic grey seals with more than 2,000 pups born every autumn. Some of the islands also support populations of rabbits that were introduced as a source of meat and have since gone wild.

The Farne Islands must be visited using local boat operators out of the nearby town of Seahouses. Local boats are licenced to land passengers on Inner Farne, Staple Island and the Longstone, depending on the time of year. There are plenty of cafes, restaurants, pubs and places to stay at Seahouses.

Grey seals

16. UPPER TEESDALE

This unique area is located upstream of the village of Langdon Beck in the north Pennines, County Durham. It encompasses an extensive upland area that includes the headwaters of the River Tees, particularly upstream of Bowlees, and consists of a diverse mix of habitats including wet heath and blanket mire as well as low intensity farmland.

Upper Teesdale is one of the most important botanical sites in the UK and has been designated an Important Plant Area. Within its 14,000 hectares, it also contains several locations that are of national importance geologically, including one of only two known outcrops of 'sugar' limestone in Britain. It is also home to several rare invertebrates and important populations of breeding waders.

Teesdale is a real jewel in the British floral crown, thanks in part to the underlying layers of Carboniferous limestone and also the legacy of the last Ice Age. The high fells are blanketed with flower-rich limestone, neutral and acid grasslands that contain species such as hoary rock-rose, Teesdale violet, bird's eye primrose, spring gentian, alpine bartsia, spring sandwort, Scottish asphodel, marsh saxifrage and Teesdale sandwort.

Teesdale violet

Plants that shelter in the rock crevices among the limestone outcrops include brittle bladder-fern, hoary whitlowgrass, alpine cinquefoil and holly fern. This area is famous for its flower-rich upland hay meadows, one of Britain's scarcest habitats. These meadows contain dozens of species per field with typical flowers such as common bistort and marsh hawk's-beard growing alongside rarer species like wood crane's bill, globeflower and a unique species of lady's mantle.

Lady's mantle

Upper Teesdale is also home to the largest juniper wood in England, and the second largest in Britain. Some of the trees that grow in the shadow of High Force are estimated to be about 250 years old.

This area has some of the highest breeding wader concentrations in mainland Britain. Curlew, lapwing and snipe all nest here in good numbers as do golden plover on the open moorland, redshank in some of the wetter areas and common sandpiper on the fast-flowing streams. It is a stronghold for England's remnant black grouse population and raven, peregrine and merlin can all be seen on the open moorland. Short-eared owl are most often seen on the moorland fringes, as are grey partridge.

With so many flowering plants in late spring and early summer, it's no surprise that on calm, warm days, the meadows are alive with bees and other insects gathering pollen. One of these is the nationally scarce moss carder bee, so-called because it gathers and combs dry vegetation to build its nest in grassy tussocks. Water voles are present in the rivers, streams and upland pools and brown hares can often be seen on the open pastures and moorland fringe.

This is largely private land but there are plenty of clearly marked footpaths in the area. Some of the walks are fairly difficult and many follow short sections of quiet country lanes. Bowlees Visitor Centre is an excellent place to start a visit to Upper Teesdale and contains plenty of information about the site as well as a café.

The B6277 runs through Upper Teesdale from Middleton-in-Teesdale to the south-east towards Alston to the north-west. There are plenty of excellent food-serving pubs in the surrounding villages and Barnard Castle on the A66 has lots of cafes and restaurants.

Low force

17. BEMPTON CLIFFS

Bempton Cliffs is a nature reserve run by the RSPB at Bempton on the Yorkshire coast. The hard chalk cliffs are relatively resistant to erosion by the North Sea storms and therefore offer sheltered headlands, ledges and crevices for nesting seabirds. The cliffs run for about 15 kilometres from Flamborough Head north towards Filey, and are some of the highest in England at over 100 metres tall.

Each spring and summer around half a million seabirds gather to nest along this short stretch of coastline, and Bempton Cliffs becomes the gateway to one of the most spectacular wildlife spectacles in the UK. Gannets return to this, their largest mainland breeding site in Britain, as early as January but nesting won't begin in earnest until late March, and by late April, 13,500 pairs will be incubating their single egg.

The gannets are joined on the nesting cliffs by thousands of guillemots, razorbills, kittiwakes, herring gulls, fulmars and shags, and although their numbers are declining globally, around 3,000 puffins also breed here in natural cracks and crevices. Most of the seabirds will have left the cliffs by early August, although the gannets will persist until late October and peregrines patrol the cliffs throughout the year in search of prey.

Gannet

Kittiewakes

Barn owls, skylarks, linnets and reed bunting all breed in the farmland along the cliff tops as do two species that have now disappeared from much of their former UK range, corn bunting and tree sparrow. In fact, the reserve is a stronghold for the tree sparrow, with around 40 breeding pairs and an overwintering population of more than 200. In winter, the resident barn owls are joined by migrant short-eared owls as they hunt small mammals over areas of rough grassland.

Its coastal location and the fact that it's on a headland means that Bempton is also an excellent site for migrant birds. Spring and autumn brings finches, warblers, chats and thrushes as well as annual rarities such as yellow-browed warblers and red-backed shrike. Indeed, just about anything could turn up so it's well worth keeping your eyes and ears open at all times.

Red-backed shrike

Harbour porpoises are often recorded offshore on calm days and in recent years, other cetaceans, including humpback whales, have been seen nearby. Early morning visits can provide sightings of roe deer and brown hare.

In summer, the reserve and cliff tops are bathed in a sea of pink as the red campion comes into full bloom. Several species of orchid have been recorded at this site, including northern marsh, pyramidal and common spotted. Many of the commoner grassland butterflies are recorded on warmer summer days, as are day-flying moths such as cinnabars, burnet moths and, occasionally, hummingbird hawk moths.

Six, safe cliff top viewpoints give spectacular views of the nesting seabirds and the dramatic coastline. The Seabird Centre at the heart of the reserve has information on all the breeding birds as well as recent sightings and volunteers are always on hand to answer questions about the reserve's wildlife. Three of the viewpoints are fully accessible with a hard surfaced path linking them in a circular route. A weekly RSPB Seabird Cruise departs from Bridlington Harbour from May to August.

The Seabird Centre has a shop, refreshments and toilet facilities and you can hire binoculars and children's activity packs throughout the year to enhance your visit.

Pyramidal orchid

Hummingbird hawk moth

18. YORKSHIRE DALES LIMESTONE IPA

Located in the north of England and covering the central Pennines in the counties of North Yorkshire and Cumbria, the Yorkshire Dales Limestone Important Plant Area is one of the botanically richest sites in the UK. Unique and breathtaking flower-rich landscapes can be found here thanks to the Great Scar Limestone laid down some 300 million years ago and today, concentrated in the uplands around Ingleborough, Malham and Wharfedale.

Limestone pavement is a rare, plant-rich habitat in Britain and the majority can be found in the western parts of North Yorkshire as well as south and east Cumbria. The Yorkshire Dales National Park contains approximately half of all Britain's remaining limestone pavement, all of which is protected by Limestone Pavement Orders. Some of the best individual sites include Gordale Scar, Kilnsey Crag, Malham Cove, Scar Close, Colt Park Wood and Sulber, although there are simply too many to mention them all individually.

The steepest cliffs in the area support plants such as common bistort and marsh hawk's-beard growing alongside rarer species like wood crane's bill. The limestone pavement with its clints and grikes (slabs and crevices) provides home for wild thyme, bloody crane's bill and common rock rose as well as scarcer species such as birds-eye primrose. The sheltered conditions in the grikes encourage hart's tongue fern, limestone fern, green spleenwort, rigid buckler fern, dog's mercury and wood sorrel as well as sanicle, enchanter's nightshade, herb Paris, angular Solomon's seal and lily-of-the-valley.

Wild thyme

Where the limestone grassland is grazed sympathetically, flowers such as bird's foot trefoil, mouse-ear hawkweed, small scabious and mountain pansy often grow in abundance whereas autumn gentian and mountain everlasting are more localised. The jewel in the Yorkshire Dales crown is undoubtedly the wonderfully exotic lady's slipper orchid. A single native plant survives on a well-protected wooded hillside, but a successful reintroduction programme means it's now growing in many of its former sites and visitors can now see these wonderful plants at Malham Tarn Field Studies Centre and Kilnsey Park Estate.

Lady's slipper orchid

Eyebright

Just as breathtaking as the limestone pavements are the Dales hay meadows set in their ancient landscape of dry stone walls and hay barns. Common meadow flowers such as buttercups, yellow rattle, eyebright, pignut and red clover grow alongside northern hawk's beard, lady's mantle, great burnet, bistort and wood crane's bill. Perhaps the finest hay meadows in Britain can be seen above the village of Muker in Swaledale.

Not surprisingly, the sympathetic farming methods that encourage such botanical diversity are also beneficial to a wide range of declining bird species. Curlew and lapwing are widespread breeders in suitable areas and despite recent declines, cuckoos and brown hares can still be seen here. Dippers, grey wagtails and common sandpipers nest along the fast-flowing streams and rivers and the Dales' clean waterways are one of the last strongholds for the native white-clawed crayfish in England.

With such a large area and so many small towns and villages in and around the Dales, there are plenty of cafes, restaurants, shops and pubs to choose from.

19. MARTIN MERE

This is one of the Wildfowl and Wetlands Trust's flagship reserves located near Burscough on the West Lancashire Coastal Plain. Designated an SSSI, a Special Protection Area and a Ramsar Site, Martin Mere was, until drainage began in the late 1600s, the largest body of fresh water in England.

Today, part of the old mere is the site of Martin Mere Wetland Centre, a 120 hectare complex of open water, marsh and grassland overlying deep peat. Like all WWT reserves, it is a mix of the wild and domestic where flamingoes can be seen alongside barn owls and pink-footed geese.

Pink-footed goose

At its best in the winter months, numbers of wintering waterbirds regularly exceed 30,000 individuals and include internationally important numbers of several species. More than 25,000 pink-footed geese overwinter on and around the reserve as do more than 2,400 whooper swans. Add more than 400 greylag geese and more than a thousand wigeon and teal into the mix and you begin to understand why around 200,000 visitors visit the reserve every year.

Whooper swan

Several thousand lapwing spend the winter months here as do hundreds of mallard, shelduck and black-tailed godwits. Good numbers of gadwall, shoveler, pintail, tufted duck, avocets and ruff can also be seen, as can hunting raptors including peregrine, merlin, hen harrier and kestrel. In spring and summer, look out for breeding avocets, tree sparrows, skylarks and Cetti's warblers as well as a few corn bunting along Fish Lane adjacent to the reserve. Water rail breed in the reed beds and bittern can be heard booming. Both barn owls and marsh harrier can often be seen hunting for prey all year round.

Stoats, weasels, water voles, foxes and otters are frequently recorded from several parts of the reserve. As a principally wetland reserve, Martin Mere has an impressive list of dragonflies and damselflies including black-tailed skimmers, brown hawkers and club-tailed dragonflies.

Nearly 150 hectares of the reserve are wet grassland meadows which hold some rare plant species such as golden dock, whorled caraway, tubular water dropwort and fine-leaved water dropwort as well as several species of orchid. Ragged robin, oxeye daisies, marsh thistles and bee orchids add colour to the reserve in summer and provide nectar and pollen for thousands of invertebrates.

In 2012, Martin Mere compiled its own Doomesday Book listing over 500 species of plant, 300 fungi, 1,500 invertebrates, nearly 300 bird species as well as 28 different mammals and 19 types of fish.

The visitor centre is impressive with a very well stocked shop and café. The reserve is open 364 days of the year, apart from Christmas Day, between 9.30am-4.30pm in winter and 9.30am-6pm in summer. Swan feeding, canoe trips and a variety of guided wildlife walks are all available at particular times of the year.

Brown hare

20. RATHLIN ISLAND

The northernmost point of Northern Ireland, Rathlin Island is found some 4 kilometres off the coast of County Antrim and has a population of around 150 people. It is 6 kilometres from east to west and 4 kilometres from north to south, is of volcanic origin and is one of 43 Special Areas of Conservation in Northern Ireland.

The island consists mainly of traditionally farmed fields with a mixture of scrub, bracken, gorse, lakes, coastal grassland and coastal heath. It is the spectacular coastline with its basalt and chalk cliffs and stacks, some up to 100 metres tall, and thousands of breeding seabirds, however, which attracts most naturalists.

Northern Ireland's largest seabird colony, with more than 100,000 birds, is located on the western edge of the island and overlooked by the RSPB managed Rathlin West Light Seabird Centre. Between April and July each year, tens of thousands of breeding guillemots, razorbills, fulmar and kittiwake compete for nest sites on the sheer cliffs whilst puffins are more commonly found on the nearby grassy slopes. Northern Ireland's only pair of breeding great skua can often be seen patrolling the cliff tops as can peregrine falcons, ravens and buzzards.

Puffin

Guillemot

Roonivoolin at the southernmost tip of the island is important for feeding chough as well as breeding lapwing and snipe. Following several years of intensive management work by the RSPB and local farmers, corncrake have returned for three out of the last four years and may well have bred after a 20-year absence. Oystercatcher and ringed plover breed along the rocky beaches.

Grey and common seals are widespread around the coast and both bottle-nosed and common dolphins are regularly seen offshore. Irish hares are observed throughout the island. Unfortunately, introduced brown rats and ferrets are also present.

Ferries to Rathlin run from Ballycastle throughout the year, weather permitting, and the Seabird Centre is open from Easter until late September. Toilets can be found inside the Seabird Centre as can some basic refreshments. The island has a range of accommodation to suit different needs, a pub, guesthouse, restaurant, community shop and gift shop. There is a bus service from the harbour to the West Light Seabird Centre.

Irish hare

21. BELFAST LOUGH RESERVE

Belfast Lough Reserve is made up of four sites, Belfast's Window on Wildlife (formerly Belfast Harbour reserve), Harbour Meadows, Holywood Banks and Whitehouse Lagoon. Whereas other reserves are famed for the beauty of their isolated location, Belfast Lough is situated within Belfast Harbour Estate and is surrounded by industry and urban development.

Only a stone's throw from George Best Belfast City Airport, these four sites are owned and managed by the RSPB and provide invaluable habitats within the city limits for people as well as wildlife. Indeed, it is their location right at the heart of Belfast and just 10 minutes from the city centre that makes them a firm favourite, especially when you consider that Belfast Lough is considered to be the richest bird reserve in the whole of Ireland.

Holywood Banks is one of the last remaining mudflats of the many that once surrounded Belfast Lough. Rich in invertebrates, it provides a winter refuge for a suite of waders and wildfowl including curlew, oystercatcher and dunlin, on their way to and from their northern breeding grounds.

Harbour Meadows is quite different to the other sites and constitutes a mosaic of habitats, including dry grassland, which is important for a wide range of invertebrates. Unlike the other three sites, however, there is no open access to Harbour Meadows and it is only open to the public on designated event days.

Oystercatcher

At low tide, Whitehouse Lagoon is a haven for hundreds of wading birds such as black-tailed godwit, redshank, oystercatcher, dunlin and curlew, all probing in the soft mud for invertebrate food. When the tide comes in, most of the birds move across the lough to Belfast's Window on Wildlife.

Belfast's WOW is the principal reserve and is the only one with a visitor centre, toilets and basic refreshments. It also has two shipping containers that have been converted into hides, giving excellent views out over the reserve whilst blending in with the industrial background. It was formed around 40 years ago when it was enclosed from the sea in order to be used as a dumping ground for silt dredged from the lough.

In late spring and summer, hundreds of pairs of common and arctic terns breed on the artificial rafts out in the lagoon as do Mediterranean gulls, and both reed bunting and sedge warblers nest in the reedbeds. The grassland areas support breeding lapwing and small numbers of black-tailed godwits in full breeding plumage can be seen en route to their Icelandic breeding grounds.

It is in autumn and winter that this reserve is at its best, particularly during high tide. More than 400 black-tailed godwits feed out on the mud flats, along with hundreds of lapwing, snipe, wigeon and teal. Curlew, oystercatcher, redshank, golden plover and dunlin all gather in good numbers and scarcer birds such as greenshank and ruff are regular autumn visitors.

Konik ponies are used to manage the reedbeds and rough grasslands and a sand martin bank and swift tower have recently been constructed to attract these declining African migrants.

Belfast WOW is open every day, except Tuesday, from 10am-4pm between November-February and between 10am-5pm between March-October. Two main entrances lead into Belfast Harbour Estate directly from the A2 and the 26/26A bus service from City Hall to Holywood Exchange runs past the reserve on weekdays.

Black tailed godwit

Reed bunting

Greenshank

22. MURLOUGH NNR

This little-known site is located in one of the most beautiful landscapes in Northern Ireland, at the edge of Dundrum Bay on the County Down coast and in the shadow of the Mourne Mountains. Owned and managed by the National Trust, it was declared Ireland's first nature reserve in 1967. This 697 acre site is a fragile, 6,000 year old sand dune system with heathland and woodland surrounded by estuary and sea.

In the twelfth century, a rabbit warren was established on the site by the Normans for their meat and pelt, the animals having had a major influence on the development of the heath and grassland ever since. A succession of severe storms in the thirteenth and fourteenth centuries resulted in a huge movement of sand which led to the unusually high dunes seen today.

Murlough offers the best and most extensive example of dune heath within the whole of Ireland and it is not surprising that such a habitat attracts a great selection of invertebrates. Of 23 butterfly species recorded on the reserve, the most notable is the marsh fritillary. With almost 800 larval webs recorded here recently, this is the biggest marsh fritillary colony in the country and one of the largest in the UK.

Other well-known species of butterfly on the reserve include small copper, common blue, dark green fritillary, grayling and the cryptic wood white. More than 750 species of moth have been recorded within the reserve boundary, more than at any other site in Northern Ireland. Highlights include small elephant hawkmoth, bordered sallow and sand dart as well as several 'firsts' for Ireland and Northern Ireland.

Important plants on the site include devil's bit scabious, the main food plant of the marsh fritillary caterpillar, bee orchid, pyramidal orchid and carline thistle. Early forget-me-not and shepherd's cress both flower in the dunes whereas wild pansy and primrose are more prominent in the grassland and in late April, the woods and large areas of the dunes are transformed by a spectacular display of native bluebells.

Breeding birds include skylark, stonechat and cuckoo but it's the autumn and winter migrants of Dundrum Inner Bay that provide the most notable ornithological spectacle. Waders such as black-tailed godwit, lapwing, curlew, oystercatcher, dunlin and redshank arrive here in the hundreds as do knot, turnstone and golden plover.

Dark green fritillary

Small copper

Marsh fritillary

Small elephant hawkmoth

Cryptic wood white

Carline thistle

Wigeon are also present in large numbers and the Bay is home to internationally important numbers of wintering pale-bellied Brent geese. During summer, large numbers of Manx shearwater, known locally as 'mackerel cocks', gather offshore while in autumn and winter, Dundrum Bay hosts several thousand common scoter that can be viewed from the high dunes.

Grey and common seals both use the sandbanks as a haul-out site and Irish stoats are frequently seen on the reserve, particularly when rabbit numbers are at their peak. Foxes and badgers are also present as are Exmoor ponies that are used to help manage the diverse habitats.

Murlough NNR is located 25 miles south of Belfast on the A24 and is only 1 mile south of Dundrum village. There are toilets, a NT car park and a visitor centre on site but they are not open all-year. The car park closes at 5pm. Newcastle, some 2 miles away along the beach, is a popular seaside resort with a selection of cafes, restaurants, shops and pubs.

23. BALLAUGH CURRAGH

Situated in the north west of the island, Ballaugh Curragh is the first Ramsar Wetland of international importance designated in the Isle of Man, and the site is also an Area of Special Scientific Interest. Its international status is a reflection of the variety of wetland habitats within its boundary, including bog pools, marshy grassland, birch woodland, modified bog and willow scrub, known locally as 'curragh'.

Encompassing the last remnants of an ancient lake that formed in the lowland between the Bride hills and the Manx upland, today the site is owned and managed by the Manx Wildlife Trust, Manx National Heritage and Curraghs Wildlife Park as well as several private landowners. Ballaugh Curragh is the largest remaining intact example of a very distinctive Manx habitat that was historically important as a plentiful source of fish, waterfowl and willow.

The Curraghs are a tapestry of scrubland, including willow and bog myrtle scrub, traditionally managed hay meadows and sphagnum-dominated bogs. These create ideal habitats for scarce plant species such as bladderwort, royal fern, lesser tussock-sedge and marsh cinquefoil. In late spring and early summer, many of the meadows are full of thousands of orchids, including common spotted orchid, heath spotted orchid, northern marsh orchid, common twayblade and greater butterfly orchid.

Ballaugh Curragh is best known for having the second largest hen harrier roost in Europe. At times, more than 60 harriers have gathered here at dusk (claims of over one hundred are almost certainly exaggerated) although recent numbers are down on this total as the Manx breeding population has declined. Other winter visitors include whooper swan, shoveler, pintail and short-eared owl as well as sparrowhawk, kestrel and merlin.

Royal fern

Marsh cinquefoil

Greater butterfly orchid

Northern marsh orchid

Common spotted orchid

Traditional management of the hay meadows has led to a recent re-colonisation by corncrakes although they are yet to become regular breeders. Curlew, lapwing and snipe also nest as do water rail, lesser redpoll and grasshopper warbler. It also holds a good population of European eels. Ballaugh Curragh also has a self-sustaining population of red-necked wallabies, a result of escapes from the Curraghs Wildlife Park in the late 1960s.

There are several access points to this site, including via the Manx Wildlife Trust reserve at Close Sartfield which is just off the Ballacrye Road heading out of Ballaugh towards Ramsey. The Curraghs Wildlife Park forms part of the area and has a café and toilets otherwise the nearby village of Ballaugh has a convenience store and a pub.

24. DEE ESTUARY

This is a huge site on the border between Wales and England that is of international significance for its populations of wintering waders and wildfowl. It is unusual in that a relatively small volume of water flows out of a large basin therefore at low tide, vast areas of invertebrate-rich sand and mud becomes accessible to the birds.

Extending from Shotton to the Point of Ayr, this is where the River Dee flows into Liverpool Bay. It has numerous reserves, managed by several different organisations, dotted along its shores but despite its high level of protection by UK and European law, the Welsh side of the Dee has been extensively industrialised.

More than 120,000 birds visit the Dee estuary every winter. Twelve species are present in internationally important numbers and a further six in nationally important numbers. Redshank, oystercatcher, knot, dunlin, black-tailed and bar-tailed godwits all gather in large flocks as do teal, shelduck and pintail. Curlew and grey plover are also present in internationally important numbers with nationally important numbers of great-crested grebe, red-breasted merganser, sanderling, cormorant and wigeon.

Dunlin

All of these birds can be seen from the Welsh side of the Dee at key sites such as Oakenholt Marsh, Flint Castle and the Point

A linnet bathes

of Ayr. Generally, birds are scattered throughout much of the estuary during low tide but form spectacular high tide roosts as the water covers the mudflats.

Large sections of the English side of the Dee have silted up to form areas of saltmarsh and rough grassland. In winter, these are home to around 6,000 pink-footed geese as well as large flocks of wintering finches such as linnet. Hunting hen harrier, marsh harrier, merlin and short-eared owl are regularly seen over the saltmarsh and peregrine can be seen wherever there are wintering waders and wildfowl.

The RSPB reserve at Burton Mere Wetlands on the Wales/England border is a focus for breeding as well as wintering birds. Avocet, lapwing, redshank and water rail all breed in good numbers and the reedbed is home to sedge warbler, reed bunting and reed warbler in spring and summer. Skylark, yellow wagtail, grasshopper warbler, lesser whitethroat and green woodpecker all nest as do little and cattle egrets in a nearby heronry. Mammals on this reserve include brown hare, otter, harvest mouse and water vole and both southern marsh and bee orchids are present, the former in large numbers.

Common terns breed on artificial islands in lagoons at Shotton Steelworks and the sheltered, shallow waters of the Dee estuary provide an important nursery for juvenile Sandwich, common and little terns. Hilbre Island near the mouth of the estuary is home to an increasing flock of brent geese as well as turnstone and purple sandpiper in winter. It is also an active bird

Sanderling

observatory and many of the estuarine waders use the island as a high tide roost.

More than 500 grey seals haul up on West Hoyle Bank near Hilbre Island although they return to the Pembrokeshire and Hebridean islands to breed. Common seals are occasionally seen in the estuary as are harbour porpoise and bottle-nosed dolphin, often following the salmon that migrate from Liverpool Bay up the river to spawn.

The Dee estuary is a large site located north west of Chester on the Wales/England border. There are too many sites around the estuary to list here but the principal ones are as follows. Point of Ayr (SJ122854); Greenfield Dock (SJ200780); Flint Castle (SJ247735); Burton Marsh (SJ301748); Parkgate (SJ275788); Thurstaston Shore (SJ240830); and Hilbre Island (SJ185880).

Purple saxifrage in rugged Cwm Idwal

25. CWM IDWAL

A dramatic amphitheatre of rock surrounding a small glacial lake in the mountains of Snowdonia, it's no wonder that Cwm Idwal has inspired generations of scientists including Charles Darwin and Sir David Attenborough. For geologists and glaciologists wishing to separate their terminal moraines from their roches moutonnées, this is the place to be, but there is plenty here to get the naturalist's heart beating too.

Declared Wales' first National Nature Reserve in 1954, Cwm Idwal is rightly famous for its scarce Arctic Alpine plants, the most famous of which is the rare Snowdon lily. In Britain, this delicate-looking plant is confined to just a handful of sites in the UK, all of them in the high mountains of north-west Wales. Flowering in May and June, at this site, it shares the mountain crags with roseroot, globeflower, lady's mantle and the unique Snowdonia hawkweed.

In early summer, mossy and starry saxifrage carpet the scree slopes around the steep path leading up the Devil's Kitchen and the patient naturalist should get good views of the sure-footed feral goats that have roamed these mountains for thousands of years. These animals are easiest to see in winter when they leave the high tops for the richer pickings in the sheltered valley bottoms and woodlands, and to give birth to their kids.

One of Snowdonia's feral goats

Bog asphodel in bloom

In late March and early April, keep your ears open for the blackbird-like call of male ring ouzels that have recently returned from their north African wintering grounds. Several raptors, including peregrine falcons, buzzards and kestrels hunt along the valley sides.

Although the glacial lake, Llyn Idwal, is not rich in bird life, great crested grebes nest amongst the emergent vegetation and both grey wagtails and common sandpipers feed along the shoreline. The lake also has a good representation of upland freshwater plants such as quillwort, water lobelia and intermediate water-starwort.

It's worth noting that Cwm Idwal is very popular with walkers, climbers and outward-bound schools and can be incredibly busy on weekends and Bank Holidays. However, the reserve is large enough to be able to escape the hustle and bustle of visitors with plenty of quiet corners to sit down and enjoy the wildlife.

Cwm Idwal is situated alongside the A5 road between Bethesda and Capel Curig, at the head of the Nant Ffrancon valley and is a short walk from the car park adjacent to the main road. (SH648604).

26. ELAN VALLEY

This is a huge site, encompassing more than 70 square miles of moorland, woodland, hay meadows, lakes, rivers and reservoirs. Within its boundary, there is a National Nature Reserve and several Sites of Special Scientific Interest and the diversity of wildlife reflects the impressive variety of habitats.

Owned by Dwr Cymru/Welsh Water and managed by the Elan Valley Trust, the area first hit the headlines when a series of reservoirs were created to provide water for the growing city of Birmingham from the 1890s onwards. Since then, most of the estate has been managed sympathetically for wildlife, principally in order to ensure high water quality.

The vast moorland surrounding the reservoirs are dominated by upland grasses and heathers with large areas of blanket bog. Hare's tail cottongrass, bog rosemary and round-leaved sundew are all widespread here. Also present are the nationally rare bog orchid and floating water plantain.

This upland habitat is the Welsh stronghold for breeding golden plover and dunlin although both are now present in small numbers. Skylark, merlin, hen harrier and short-eared owls can also be seen on the high tops as can foraging red kite, buzzard, raven and kestrel. Whinchat breed on some of the bracken-dominated slopes where cuckoos can still be heard.

The Elan Valley Estate boasts some of the finest upland hay meadows in Wales. Sensitive management means that once-common species such as great burnet, mountain pansy, wood-bitter vetch, fragrant and greater butterfly orchids are all doing well at these sites. The meadows also contain pignut, bluebells and wood anemones as well as a wide range of commoner species.

Several of the steeper slopes around the reservoirs are covered in old sessile oak woodlands, a wonderful habitat for mosses, liverworts and lichens. The woodlands also provide a home for breeding redstarts, wood warblers and pied flycatchers and both hawfinch and lesser-spotted woodpecker have been seen at the adjacent RSPB Carngafallt reserve near Elan Village.

The mature coniferous woodlands are also worth a look as crossbill, siskin, sparrowhawk and goshawk are regularly seen here. The reservoirs themselves are too deep to be of any great conservation value but goosander, tufted duck, great-crested grebe and teal are common and grey wagtail, black-headed gull and common sandpiper all breed.

Otters are occasionally seen on the lakes and rivers, hares are widespread but elusive on the moorland edges and common lizards can be seen basking on rocky outcrops and stone walls on sunny summer days. Twenty seven species of butterflies have been recorded, including the purple hairstreak, and amongst the 200 plus species of moths seen on the estate is the rare Welsh clearwing which favours old birch woodlands.

Paths, cycle paths and tracks are plentiful and a large visitor centre, café, toilets and shop is situated below the Caban Coch Dam. The Elan Valley Estate has been declared an International Dark Sky Park so a night visit is also well worthwhile.

From the town of Rhayader, turn west off the A470 and onto the B4518. Follow this road for 4 kilometres. The visitor centre is below the first dam (SN928646).

A kestrel grips its prey

A goshawk

27. SKOMER AND SKOKHOLM

Skomer National Nature Reserve is the wildlife jewel in the Welsh crown. It can hold its own with the best the world has to offer and a visit to the island is a must for anyone interested in wildlife. Bought in 1959, the island is managed by the Wildlife Trust of South and West Wales.

Skomer is situated about 1.5 kilometres off the south Pembrokeshire coast near the small village of Marloes and the absence of ground predators such as rats, cats and foxes means that it is a haven for wildlife. Famed for its hundreds of thousands of seabirds, Skomer has a great deal to offer both above and below the waves. In spring and early summer, it is resplendent under a carpet of bluebells and red campion as well as thrift and sea campion. Later in the year, the heather turns parts of the island purple and even in winter, yellow gorse flowers add a splash of colour.

This island is famous world-wide for its breeding seabird colonies. Visitors between late April and early July can expect to see thousands of breeding puffins, razorbills, guillemots and four species of gulls. Fulmar, peregrine falcon, kestrel, chough and raven also nest on the sea cliffs and up to ten pairs of short-eared owl nest among the taller vegetation.

Manx shearwater

Skomer's star bird is the Manx shearwater, with hundreds of thousands on the island. By day, however, they are either in their nesting burrows or far offshore searching for food, returning to the island only after dark in order to avoid the large, predatory

The sea slug *coryphella lineata*

Pink sea fan

The bryozoan, *Pentapora foliacea*

gulls. Visitors who stay on the island overnight will enjoy both the spectacle of tens of thousands of shearwaters flying above their heads and hundreds of common toads crawling around their feet, searching for slugs. Other breeding birds include skylark, reed bunting and oystercatchers, and due to its coastal position, several rare migrants are recorded annually.

Slow worms can be seen around the old farmhouse in the centre of the island and the unique Skomer vole is common but rather elusive. After dark in early summer, glow-worm females can be seen attracting males with their green lights, oxidizing a compound called luciferin. Grey seals are common throughout the year and haul out onto the beaches to give birth in early autumn. Harbour porpoises and several species of dolphins are regularly recorded offshore and as Wales' only Marine Conservation Zone, Skomer attracts divers from all over the world to see species such as the pink sea fan, rose coral and several rare sea slugs.

Skokholm is Skomer's quieter cousin, just as rich in wildlife but with far fewer visitors. Just south of its more famous neighbour and 4 kilometres off the coast of Pembrokeshire, Skokholm became Britain's first bird observatory in 1933 under the guidance of its warden, the naturalist and author Ronald Lockley. After an absence of almost forty years, the observatory has re-opened and attracts bird ringers from around the world. Managed by the Wildlife Trust of South and West Wales for fifty years, they were able to buy the island in 2006 and it was designated a National Nature Reserve in 2008.

Due to its isolated nature and the clean, damp air from the Atlantic, Skokholm is home to several nationally scarce species of lichens, including the golden hair lichen. The invertebrate fauna is also exciting, containing nationally rare species of moths such as the black banded and Devonshire wainscot. Botanically, it is made up of maritime grassland with some wet heath and saltmarsh housing species such as thrift, three-lobed water crowfoot, marsh St John's wort and small nettle.

From mid-March onwards, Manx shearwaters return to the island to nest in underground burrows. With 45,000 breeding pairs, Skokholm is the third largest colony in the world and the island also supports up to 5% of the global population of storm petrels, small, nocturnal, bat-like seabirds that nest in crevices in stone walls and steep scree slopes. An estimated 3,500 pairs add their eerie calls to those of the shearwaters on dark summer nights.

Up to 6500 puffins nest in old rabbit burrows and both razorbills and guillemots are common on the sea cliffs. The fulmar, chough, oystercatcher, and wheatear populations have all been studied intensively as have the large colonies of lesser black-backed, herring and great black-backed gulls.

Its position off the coast means that in spring and autumn, Skokholm attracts thousands of passage migrants, principally warblers, thrushes, pipits and flycatchers that nest in Europe. Scarcer visitors such as wryneck, golden oriole and hoopoe are recorded annually and sometimes exceptional rarities are noted, like Baltimore oriole, Swainson's thrush and scops owl.

The island is within the boundary of the Marine Conservation Zone and the seabed is rich in wildlife. Several species of cetacean are regularly recorded including harbor porpoise, common dolphin, Risso's dolphin and, occasionally, larger and scarcer species such as minke whale.

Getting there – From Haverfordwest, take the B4327 towards Dale then follow the signs towards Marloes and Martin's Haven, from where the boats leave. The car park is owned by the National Trust and is free to members only. Because they are islands, visiting is not easy. Steep steps mean that unfortunately, it is not accessible to wheelchair and pram users and fees have to be paid for the boat and for landing. Boats for Skomer depart daily from Martin's Haven at 10am, 11am and 12pm from 1st April until the end of September, apart from Mondays or during bad weather. There are no day visits to Skokholm, making it a really special place to spend a long weekend with only a handful of like-minded people. Boats run on Mondays and Fridays and to avoid disappointment, bookings should be made well in advance. Tickets for the Skomer daily cruise can be bought at the Wildlife Trust's Lockley Lodge at Martin's Haven. Overnight bookings for either island should be made directly from the Wildlife Trust of South and West Wales.

Storm petrel

Wryneck

Skokholm cliffs

28. COED-Y-BWL

This is yet another of Wales' hidden little gems, a deciduous woodland tucked away in a river valley in the Vale of Glamorgan. Coed-y-Bwl is an ancient ash woodland situated on the northwest side of the Alun valley and is managed under lease by the Wildlife Trust of South and West Wales.

The southern end of the wood was formerly dominated by elm but the trees were devastated by Dutch elm disease more than thirty years ago. Since then, this area has been clear felled and replanted with ash, lime and wild cherry. The northern end is made up of ash and sycamore with an understory of field maple.

This reserve is best known for its wild daffodils that carpet the southern part of the wood in a sea of yellow in March and April every year. Once common in woodlands and hedgerows throughout south Wales, wild daffodils are now a rare sight in the Welsh countryside, having been replaced in most areas by non-native imports. Interspersed between the daffodils in early spring are the white flowers of wood anemones whilst the northern slopes are dominated by bluebells.

Breeding birds at this site include several warblers such as blackcap and willow warbler as well as woodland specialists such as treecreeper and nuthatch. Marsh tit, a scarce and declining species throughout Wales, can be seen here and the wood is a good place to see raptors such as sparrowhawk and, at dusk, tawny owl.

Fungi include the jelly ear that grows on elder trees and the colourful scarlet elf cup that can be seen growing on decaying branches on the woodland floor in early spring. The list of mammals seen on the reserve includes fox, badger and stoat.

A well-defined footpath leads through the southern end of the reserve and gives excellent views of the flowering daffodils. It is muddy and steep in places, however and is unsuitable for wheelchair access but the flowers can be seen from the road.

This is a difficult reserve to locate but it's well worth making the effort. From the A48 south of Bridgend, turn left onto the B4265 towards Ewenny. Here, turn left down Wick Road and stay on this road (forking right on two occasions). After about a kilometer, take a right turn down Heol y Stepsau, cross a ford and continue for about 600 metres. The reserve is on your right. (SS909749)

Young tawny owls

Fox cub

Scarlet elf cup

29. GWENT LEVELS

This large reserve was created in 2000 as mitigation for the destruction of the vast areas of intertidal mudflats when the Cardiff Bay Barrage was built. It is owned and managed by Natural Resources Wales in partnership with the RSPB and Newport City Council and others, and covers 1,080 acres of wet grassland, grazed pasture, reedbed and shallow pools on the north shore of the Severn Estuary east of Newport, as well as 1,000 acres of inter-tidal mudflats leased from the Crown Estates.

The numerous ditches, known as reens, are particularly rich in plants including several scarce species such as rootless duckweed, the smallest flowering plant in the world, frog-bit and flowering rush. Reedmace beds are widespread and the localized celery-leaved buttercup flowers between May and September.

Reedmace

In May and June, it's worthwhile following the 'Orchid Trail' from the visitor centre. On the way, you will encounter hundreds of orchids from five different species, namely southern marsh, bee, pyramidal and common spotted orchids as well as marsh helleborine.

This reserve is also noted for its large numbers of scarce invertebrates. These include great silver water beetle, shrill carder bee and brown-banded carder bee. Ruddy darters and hairy dragonflies take advantage of the numerous pools and ditches, and the wildflowers attract good numbers of butterflies such as ringlet and common blue, and colourful day-flying moths such as scarlet tiger and narrow-bordered five spot burnet.

Common blue butterfly

Water voles, recently re-introduced at the Gwent Wildlife Trust's Magor Marsh reserve, have now colonized this site. Weasels are often seen running across the paths and otters are usually encountered as they hunt for rudd in the reedbeds.

This National Nature Reserve is important for a number of breeding bird species, particularly waders, and the shallow scrapes hold the most important population of nesting avocets in Wales. Lapwing, redshank and little-ringed plover also breed here, as do waterfowl such as gadwall, tufted duck and great-crested grebe.

Newport Wetlands' extensive reedbeds are home to nesting bearded tits, water rail and Cetti's warblers and in winter, tens of thousands of starlings gather here to roost. Winter is also the time when hundreds of wigeon, shoveler, lapwing and black-tailed godwits gather to feed on the flooded grasslands and large flocks

Starlings

of dunlin, shelduck, redshank and curlew feed along the mudflats and saltmarsh. The cryptically-coloured and elusive bittern is another regular winter visitor.

There are plenty of good quality footpaths on the reserve, nearly all suitable for pram and wheelchair users and three mobility scooters as well as two wheelchairs are available to loan free of charge from the visitor centre. A floating pontoon pathway leads through one of the reedbeds towards the lighthouse and a cycleway follows the north shore of the Severn. The visitor centre contains a well-stocked shop, including a wide range of binoculars, telescopes and books, and an excellent café.

The entrance to Newport Wetlands is off West Nash Road, between the village of Nash and Uskmouth power station. Take junction 24 off the M4 and follow the A48 south. At the 4th roundabout, take the Nash Road exit heading east and follow the brown duck signs.

30. RUTLAND WATER

Set in the heart of the East Midlands, Rutland Water is one of the largest man-made lakes in Europe. Completed in 1975 and officially opened the following year, the reservoir is filled by pumping water from the River Nene and River Welland and covers 10.86 square kilometres when full. Owned by Anglian Water, it provides water to the East Midlands, including the city of Peterborough.

The reservoir is used not just for water storage but also for recreation including sailing, wind surfing, water skiing and fishing, and a 37 kilometre perimeter track is very popular with cyclists, walkers and runners. A 1,555 hectare area of lake and shore has been designated a Special Protection Area, a RAMSAR site and a Site of Special Scientific Interest, and 393 hectares at the western end is managed as a reserve by the Leicestershire and Rutland Wildlife Trust.

More than 25,000 waterfowl regularly spend the winter on the reserve, including peak numbers of more than 9,700 tufted duck, over 4,000 wigeon and more than 4,000 coot. Shoveler and gadwall are present in internationally important numbers and are joined each autumn and winter by hundreds of pintail, teal and mallard as well as goldeneye and goosander, and smew numbers often reach double figures.

Shoveler

Golden plover and lapwing gather in large numbers on the islands, periodically put to flight by hunting peregrine falcon. At dusk, short-eared and barn owls can often be seen hunting small mammals over the meadows near the Anglian Water

Water rail

Birdwatching Centre and bitterns hide away in the reedbeds.

Shelduck, teal, mallard, pochard, gadwall, shoveler and tufted duck all breed in the vegetated fringes of the reservoir, as do several pairs of water rail, and both sand martins and kingfishers nest in man-made banks. Great-crested grebes breed in good numbers and the muddy shorelines and islands are perfect for nesting little ringed plover and lapwing. Around 60 pairs of common terns nest, most on the floating platforms and islands where they are able to escape from one of their principal predator, otters! More than 50 pairs of cormorant nest here and a heronry of up to 17 nests is located in the mature trees near the fish ponds.

Several species of warbler, such as reed, sedge and grasshopper, breed on the reserve as do a few pairs of nightingale but in recent years, Rutland's main attraction has been several pairs of breeding ospreys. First re-introduced to this site in 1996, ospreys have been breeding successfully since 2001, the first nesting in England for over 150 years, and the success has continued with 8 pairs rearing 16 young in 2017.

The lake is stocked with brown and rainbow trout, but coarse fish have been pumped in along with the water from the rivers Welland and Nene. These include roach, bream, zander, pike, eel, perch and carp. Up to 24 species of butterfly and more than 500 species of moths have been recorded, including notable species such as red-tipped clearwing, angle-striped sallow and the concolorous.

Concolorous moth

In spring, the herb-rich meadows are a sea of colour with cuckooflower, ragged robin and yellow rattle to the fore. Water violets are common in the shallow waters and are often used as perching posts by dragonflies and damselflies such as the black-tailed skimmer and emerald damselfly. Otters are present, but extremely elusive, whereas stoats, weasels and muntjac deer are often seen crossing paths or from the numerous hides scattered around the reservoir.

Since 1989, Rutland Water has been home to the British Birdwatching Fair. Described as the naturalist's Glastonbury, it is an annual event for birdwatchers and naturalists that attracts suppliers of optics, books, clothing, holidays, artwork and just about anything a wildlife enthusiast could want. It also has an impressive list of speakers and attracts tens of thousands of visitors for one weekend in August. The Birdfair has now raised and donated more than five million pounds for international conservation projects. With its two visitor centres, more than 30 birdwatching hides and central location, Rutland Water is the ideal venue for such an event.

31. THE WASH

The Wash is a 62,046 hectare rectangular bay and estuary at the north-west corner of East Anglia, on the border between Norfolk and Lincolnshire. One of the broadest estuaries in the United Kingdom, it is fed by the rivers Witham, Welland, Nene and Great Ouse, and is one of the most important sites in Europe for wintering waders and wildfowl.

Much of the Wash is very shallow with numerous large mud and sandbanks that are exposed at low tide, particularly in the southern Wash. Due to sedimentation and land reclamation, the coastline has altered significantly over the centuries and historically, several towns once on the coast are now some distance inland. It is made up of very extensive salt marshes, large intertidal sand and mud banks, shallow waters and deep channels.

It is estimated that about 2,000,000 birds a year use the Wash for feeding and roosting during their annual migration with an average total of around 400,000 birds present at any one time. It is internationally important for 17 species of birds, namely pink-footed goose, dark-bellied brent goose, shelduck, pintail, oystercatcher, ringed plover, grey plover, golden plover, lapwing, knot, sanderling, dunlin, black-tailed godwit, bar-tailed godwit, curlew, redshank and turnstone.

Flock of wigeon

Turnstone

Grey plover

This site is at its best during the winter months, particularly on a rising tide between September and April. Thousands of wading birds take off from the mudflats as the tide encroaches, turning and tumbling in one of the UK's most incredible wildlife spectacles. This is best viewed at either RSPB Snettisham or Gibraltar Point NNR.

It is difficult to choose just a handful of locations within such an important site but there are some excellent nature reserves scattered along the boundary of the Wash. The RSPB's Snettisham reserve, with numerous hides overlooking saline lagoons and mudflats, is a great place to watch high tide roosts of knot and godwits as well as pink-footed geese as they make their way to and from roost in their thousands at first and last light.

RSPB Frampton Marsh is one of the best places to watch waders in the UK with over 34 species recorded annually. Recently, it held 258 curlew sandpipers, the largest flock in the UK in recent times. In autumn, up to 7,000 black-tailed godwits can be seen and in winter, up to 20,000 mixed waterbirds can be seen on the wetland areas on a daily basis. These large gatherings of birds often attract the attention of predators such as peregrine, merlin and hen harrier as well as the resident marsh harriers.

Gibraltar Point NNR, at the northern extremity of the Wash, supports important wintering populations of waders such as grey plover, sanderling, bar-tailed godwits and knot. In summer, little tern nest along the sandy and shingle beaches and turtle doves, now a very scarce breeding bird in the UK, still nest in small numbers on the reserve. Gibraltar Point is also home to one of the UK's 19 bird observatories where local and visiting ornithologists collect valuable information about the movements, condition and populations of visiting birds.

The Wash is an important breeding and feeding area for both grey and common seals, both of which can be seen hauled out on the many sandbanks scattered throughout the site.

There is a café and shop with excellent panoramic views at Gibraltar Point NNR, a delightful visitor centre with catering facilities at RSPB Frampton Marsh and an excellent visitor centre, shop and café at RSPB Titchwell Marsh on the nearby north Norfolk coast. Several of the surrounding towns and villages have excellent pubs and cafes.

RSPB Snettisham

32. RSPB MINSMERE

Described by Chris Packham as the "Eurodisney of British reserves', Minsmere certainly delivers when it comes to wildlife. Owned and managed by the RSPB since 1947, the reserve lies within the Suffolk Coast and Heaths Area of Outstanding Natural Beauty and its designation as an SSSI, an SPA, an SAC and a Ramsar Site gives some indication of its importance for wildlife. It's no wonder that more than 5,800 species have been recorded here.

Within its boundaries, the 1,000 hectare site contains a wide range of habitats including reed bed, lowland heath, acid grassland, wet grassland, woodland, coastal lagoons and shingle vegetation, all managed for wildlife. This haven began its life, however, as a re-flooded area behind Suffolk's coastal dunes to defend against German invasion during the Second World War!

Minsmere's reedbeds are home to scarce birds such as bittern, marsh harrier and bearded tit as well as water rail, reed and sedge warbler. Water voles, water shrews, grass snakes and otters also reside here, the latter taking advantage of the 13 species of fish found on the reserve. Several dragonflies and damselflies fly over the reed vegetation and open water on warm summer days, including the rare Norfolk Hawker, an East Anglian speciality.

Norfolk hawker
dragonfly

The heathlands are at their best in spring and summer when stonechat, woodlark and Dartford warbler share songposts with churring nightjar. Adders and common lizards often bask on sheltered, south-facing slopes and although natterjack toads breed in shallow pools nearby, there is no public access due to the sensitive nature of the habitat.

Acid grassland areas are grazed by herbivores such as rabbits and red deer, providing ideal breeding habitat for the rare stone curlew. This unusual-looking wading bird can be surprisingly elusive but RSPB staff are usually on call to help visitors spot them. Konik Polski ponies are used to manage the lowland wet grasslands for breeding waders such as lapwing and redshank.

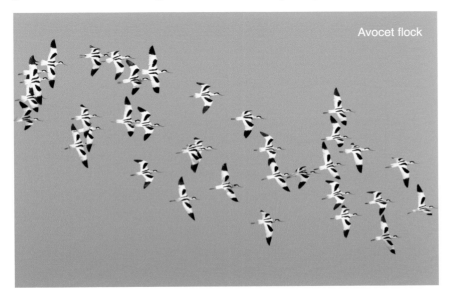

Avocet flock

Minsmere's famous scrapes host over 100 breeding pairs of avocets as well as more than 40 pairs of Mediterranean gulls and both nightingale and turtle dove can be heard singing from some of the scrubby woodland areas. In summer, hobbies can be seen hawking for insects virtually anywhere on the reserve and the repetitive call of the cuckoo, now a very scarce sound over most of the British countryside, is a familiar Minsmere backdrop in late spring.

Over 1,000 species of butterflies and moths have been recorded on the reserve including a large breeding population of the silver-studded blue as well as flame wainscot, Fenn's wainscot and white-mantled wainscot. Two species of invertebrates that are currently expanding in range, the European beewolf and the antlion, are found here as is the hairy-legged mining bee.

Silver studded blue

Dead and decaying trees in the woodlands support more than 1500 species of fungus, including rarities like moor club, deceiving bolete and lion's mane mushroom. The shingle ridges on the beaches support a variety of uncommon plants including yellow horned poppy, whereas red-tipped cudweed occurs close to Bittern Hide and round-leaved wintergreen was recently rediscovered in the reedbed near Island Mere.

Minsmere is very well equipped with an excellent café, shop and visitor centre as well as 8 hides and a network of trails. The reserve is located at the end of Sheepwash Lane and is signposted with brown tourist signs from the nearby village of Westleton.

Lion's mane mushroom

33. WHITTINGTON LODGE FARM

It might seem unusual to include a working farm in a list of 40 top wildlife sites but Whittington Lodge Farm is a revelation. A 280 hectare mixed farm situated on thin Cotswold soil near the town of Cheltenham in Gloucestershire, it has been farmed by Ian Boyd for more than 40 years.

Having grown monoculture cereals and watched the wildlife decline significantly for the first twenty-five years, Ian decided on a radical change of direction, helped by the Environmental Stewardship schemes. Extensive wildflower meadows were created and herbal leys were introduced as part of the arable rotation. A herd of native Hereford cows and calves were brought in to manage the grassland, leys and meadows. A combination of organic, regenerative agricultural and pasture-for-life farming practices have been used to bring life back to the soil which is, in turn, bringing wildlife back to the farm.

In early summer, the wildflower meadows are a sea of pyramidal, spotted and occasional bee orchids with cowslips, yellow rattle, oxeye daisies, sainfoin and many more. The herbal leys are sown with 15 different types of grasses, legumes and herbs and they all flower and set seed before the migrating herd of cows graze the field.

All these flowers attract an impressive array of bees and butterflies including common, small and holly blue, marbled white, dingy skipper and a colony of duke of Burgundy fritillaries. Skylark nest in abundance and snipe are regular winter visitors because of the high earthworm numbers.

Duke of Burgundy fritillary

Grass margins and wide conservation headlands around the cereal fields provide nesting sites for grey partridge, yellowhammer and corn bunting as well as refuge for brown hare and roe deer, both of which venture out to feed in good numbers in the early mornings and late evenings. Many rarer arable flowers grow here including cornflowers, corn marigold, corncockle, venus looking-glass and pheasant eye. The plentiful supply of seed-bearing plants also attracts large flocks of goldfinch, linnet and reed bunting.

A series of pools created for wildlife are now home to great crested newts as well as grass snakes, frogs and common toads, and slow worms find refuge in many of the areas left fallow for wildlife. An abundance of mice and voles has encouraged kestrels and barn owl to breed and hobbies hunt hirundines and large insects around the pools. Lesser horeshoe bats hunt along the mature hedgerows and woodland edges at dusk and short-eared owl, hen harrier and merlin are often recorded in winter.

Sainfoin in the orchard

Brown hares

Groups of visitors are welcome to visit the farm by prior arrangement and farm tours are organized for schools and special interest groups. There is a 5 kilometre circular walk on footpaths around the farm starting at Whittington church. Although there is no café on site, the farm does sell its own special 100% grass-fed beef and there are plenty of places to stay and eat in the nearby town of Cheltenham.

Common toad

34 LONDON WETLAND CENTRE

It's not often you find a flagship nature reserve on the edge of a capital city with 14 million inhabitants, but that is exactly what the Wildfowl and Wetlands Trust has created on the outskirts of west London. Located in the borough of Richmond upon Thames, the site is formed of four disused Victorian reservoirs tucked into a loop on the south bank of the River Thames.

The Centre, which opened in 2000, occupies more than 40 hectares of land, much of which has now been designated a Site of Special Scientific Interest. It has been carefully managed to produce a variety of wetland habitats including lakes, pools and wet grasslands, all of which attract an incredible array of wildlife, some of which cannot be seen anywhere else in London.

Like all the WWT's nature reserves, the London Wetland Centre has a collection of waterfowl from all corners of the globe but it also has an impressive list of wild British birds. Kingfishers and sand martins breed here as do water rail, an elusive species that is scarce throughout its range in the UK. A camera placed in the sand martin nesting bank relays live pictures for viewing by the visiting public.

Cetti's, sedge and garden warbler all breed as do blackcap, whitethroat and lesser whitethroat. Common terns nest on specially designed tern rafts and a variety of wildfowl, including gadwall, shoveler, pochard and tufted duck nest in the waterside vegetation. Lapwing and little-ringed plover can be seen on the more open areas whilst both great-crested and little grebe build floating nests on the open water.

Great crested grebe

Outside the breeding season, passage migrants such as green sandpiper and greenshank feed along the shores of the shallow pools. Wintering numbers of gadwall and shoveler reach nationally significant levels and this is also the best time of year to look for visiting bittern and water rail in the denser reedbeds.

Water voles flourish in the safe environment provided by this reserve as do marsh frogs, introduced into the UK in the 1930s. Common lizard, grass snake and slow worms can all be seen basking, particularly in the early spring sunshine. Summer sees an amazing display of snake's head fritillaries in the meadows, often alongside an impressive number of southern marsh orchids with a few pyramidal and bee orchids amongst them.

Meanwhile, the night skies become the domain of nine species of bat, including the nationally rare Nathusius' pipistrelle and Leisler's bats. The thriving hedgehog population also features prominently in night-time safaris that visitors can participate in on the reserve.

30 species of butterflies and more than 650 species of moths have been recorded here along with 26 different dragonflies and damselflies, including the rare hairy dragonfly. Recent invertebrate surveys have demonstrated an impressive assemblage of other invertebrate groups such as ants, bees and wasps, including the orchard bee, a species new to Britain. More than 250 species of fungi also occur, including a nationally rare brittlegill, Russula atrorubens.

The London Wetland Centre has 6 hides as well as a well-stocked shop and café. All the paths are flat and accessible to all.

Snake's head fritillary

Bee orchid

Southern Marsh orchid

35. DUNGENESS NNR

Located at the tip of the Romney Marsh peninsula in Kent, Dungeness is a unique landscape in British terms. It has no boundaries yet its desolate appearance belies a wealth of wildlife in one of the largest shingle landscapes in the world. The complex landform and sheer size of Dungeness make it one of the best global examples of a shingle beach.

Jointly owned and managed by Natural England, EDF and the RSPB with support from the Romney Marsh Countryside Project, this National Nature Reserve represents the most diverse and extensive example of stable vegetated shingle in Europe. It is of international conservation importance for its geomorphology as well as its plant and invertebrate communities and its birdlife.

Although famed for its vast area of shingle beach, the site also supports a wide range of other habitats including freshwater pits, reedbeds, wet grassland and wildflower meadows, making it one of the most species-rich wildlife sites in the country.

Dungeness is home to more than 600 species of plants, a third of all plants found in the UK. The shingle beaches are home to the critically endangered red hemp-nettle as well as other rare species such as Nottingham catch-fly, yellow vetch, lizard orchids and one

Lizard orchid

Early spider orchid

small colony of the early spider-orchid. Jersey cudweed grows on the margins of gravel pits in Dungeness RSPB reserve and nearer the shore specialized pioneer plants such as sea kale and sea cabbage grow alongside yellow horned-poppy and viper's bugloss.

Invertebrates also thrive here, notably moths, with an incredible 692 species recorded to date. This is the only site in Britain for the Sussex emerald moth whose caterpillars feed on wild carrot and several other rarities including bright wave, ground lackey, white spot, marsh mallow moth and the pigmy footman are also found here. The shingle ridges are home to an endemic leafhopper and the short-haired bumblebee, a species declared extinct in Britain in 2000, has recently been reintroduced to the wildflower meadows on the RSPB reserve in partnership with the Bumblebee Conservation Trust. Two scarce species of carder bee, the red-shanked and the brown-banded are also found here.

The gravel pits and pools are home to thriving populations of great crested newts and medicinal leeches as well as breeding common terns on some of the artificial islands. Little-ringed plover nest on the open shingle and both reed and sedge warblers breed in the reedbeds as do bittern, marsh harrier and bearded tits. Cuckoo and yellow wagtail nest here and the wet grasslands are home to breeding lapwing and redshank.

When it comes to birds, Dungeness is perhaps better known for its migrant and wintering populations. Warblers, chats and hirundines all pass through in good numbers in spring and autumn as do waders, finches and thrushes. In winter, large numbers of wigeon, teal, gadwall, pintail, shoveler, pochard and tufted duck gather on the gravel pits and this is one of the best sites in the country for wintering smew.

To reach the National Nature Reserve, follow the Dungeness Road from the nearby village of Lydd. The RSPB reserve has a visitor centre with toilets, refreshments and a shop and there are several cafés and restaurants nearby including the excellent Dungeness Snack Shack.

Smew

Sedge warbler

36. KINGLEY VALE NNR

K ingley Vale is a remarkable place that boasts one of the finest yew forests in western Europe. Tucked away in a fold of the South Downs National Park near Chichester, it is owned by Natural England and the West Dean Estate and was designated one of the country's first National Nature Reserves in 1952. In addition to its natural history, the reserve also contains one of the most important concentrations of well-preserved archaeological sites in southern England, including fourteen scheduled monuments.

This site is renowned for its twisted and ancient yew trees and includes a grove of veteran trees that are at least 500 years old, with the oldest measuring more than 5 metres in girth. Their survival is notable because most ancient yew trees across Europe were felled after the fourteenth century in order to make the staves of English longbows. The yew woodland is the principal reason for the site's designation as a Special Area of Conservation under European law.

Although the main attraction is the ancient yew forest, Kingley Vale has much more to offer. Surrounding the woodland is some superb chalk grassland with its associated rich botanical and invertebrate communities. Several species of orchid can be seen here in early summer, including bee, frog and fly orchids, as well as commoner chalk-loving plants such as birds foot trefoil and fairy flax. Several uncommon species are also present, including autumn gentian.

Frog orchid

Holly blue

The wide range of flowers attracts a variety of butterflies including holly blue, brimstone and the scarce chalkhill blue. Indeed, of the 58 species of butterfly that breed in England, 39 have been recorded at Kingley Vale. Other notable species include purple hairstreak, brown hairstreak and grizzled skipper. Grazing by belted Galloway cattle and herdwick sheep has greatly improved the grasslands for many of the butterflies as well as other invertebrates.

Grizzled skipper

Chalkhill blue

The presence of broadleaved trees such as ash, oak and elder as well as calcareous scrub with juniper, hawthorn and dogwood means that the reserve supports a rich community of breeding birds. Green woodpecker, woodcock and marsh tit all nest and raptors such as red kite, sparrowhawk and hobby are regularly seen hunting. In recent years, turtle doves have been recorded calling throughout the summer but breeding has not yet been proven. Mammals include roe deer, yellow-necked mouse, water shrew and dormouse and both badger and fox are known to use the woodland for shelter and breeding.

Kingley Vale is a magical site but the terrain is steep and uneven and much of the area is therefore not suitable for disabled access. However, wheelchairs can access the ancient yew grove at the valley bottom except in especially muddy conditions. There are two car parks that allow access to the site; one is located on the outskirts of the village of West Stoke, the other is along a minor road leading north eastwards from Stoughton. There are no on-site facilities but the town of Chichester, 5km to the south east, has plenty of cafés and pubs.

Fly orchid

37. RSPB ARNE

One of the RSPB's flagship reserves, Arne occupies a peninsula in the Isle of Purbeck in Dorset that protrudes into Poole Harbour. The charity first bought land here in 1965 to safeguard rare heathland species but it has now grown to more than 835 hectares in size and encompasses a wide variety of habitats including oak woodland, farmland, saltmarsh, mudflats and reedbed.

Arne remains one of the few areas where all six of Britain's native reptiles can be found. The rare smooth snake is widespread here, but elusive, as is the adder. Sand lizards, however, can be seen basking on warm sunny days at certain heathland locations adjacent to the numerous trails, as can common lizards and slow worms. Grass snakes prefer areas adjacent to the various freshwater pools and ponds where they can hunt for frogs and toads.

One of the largest spiders in the UK, the raft spider, is found here as is the impressive wasp spider, and in 2011, the extremely rare ladybird spider was introduced to the reserve. With its shallow pools, Arne is a haven for dragonflies and damselflies, with 28 species recorded on the reserve to date, including the small red damselfly and the downy emerald. It also boasts over 600 species of moths and good numbers of butterflies with a particularly strong population of silver-studded blues as well as grayling and woodland specialists such as purple hairstreak.

Wasp spider

Green woodpecker

Sika deer

In spring and summer, the open heathland is home to 70 pairs of Dartford warbler as well as woodlark and hobby. Stonechat, green woodpecker and cuckoo can also be seen as can the heath tiger beetle and the Dorset heath, a defining species of the area's heath and bogs. A summer dusk visit in good weather could be rewarded with 'churring' nightjars, and possibly displaying males.

The sika deer rutting season peaks in October and November, the piercing calls of the stags echoing across the reserve. Autumn is also a good time to look out for migrating ospreys in Poole Harbour although a recent re-introduction scheme means that these elegant birds of prey can be seen here throughout the spring and summer months and it can only be a matter of time before they stay to breed.

Large numbers of waders such as avocets, black-tailed godwits and dunlin gather to feed on the mudflats in the winter months and Poole Harbour hosts the largest flock of spoonbill in the UK with 75 birds recorded in the autumn recently. Hundreds of dark-bellied brent geese are joined by wildfowl such as wigeon, teal and pintail. Hen and marsh harriers regularly hunt over the reedbeds and saltmarsh in winter and the thousands of waders and wildfowl attract hunting peregrine falcon and merlin.

Due to its stunning views across Poole Harbour, the wildlife spectacles and ample facilities, Arne is a popular destination for visitors. The shop and café, opened in 2016, are an excellent addition and the nearby town of Wareham has plenty of restaurants and pubs.

Spoonbills

38. SLIMBRIDGE WWT

One of the most famous nature reserves in the UK, Slimbridge was set up by the artist and naturalist Sir Peter Scott in November 1946. Conveniently situated on the eastern bank of the Severn Estuary midway between Bristol and Gloucester, it is a magnet for wintering wildfowl and waders from the arctic tundra.

The reserve comprises 800 hectares of saltmarsh, wet pasture, reed bed and lagoons. A small part of land is encompassed by a fox-proof fence to guard the extensive collection of waterfowl from all over the world. Outside the fence, however, lies a wild area of land managed specifically for wildlife.

In winter, Slimbridge is home to around 30,000 wildfowl and waders. Thousands of lapwing, golden plover, dunlin and wigeon are joined by hundreds of curlew and European white-fronted geese, although numbers of the latter are significantly reduced from the peak of 7,000 recorded during the 1940s. Since the early 1950s, the Bewick's swans have been fed on The Rushy lake and throughout the winter, the public are able to enjoy a daily commentated wild bird feed.

Bewick's swan

Pintail, pochard, tufted duck, gadwall and teal gather during feeding times at The Rushy, but also on the flooded fields. Spotted redshank, little stint and ruff can also be seen in good numbers and the newly established population of breeding common cranes can be seen here throughout the year. Starlings and jackdaws often gather to roost in huge numbers at dusk and this large flock inevitably attracts hunting sparrowhawk, peregrine falcon and even goshawk.

During the breeding season, the wet grassland areas are home to nesting lapwing and redshank, and avocets now nest on the shallow pools. Kingfishers nest adjacent to the aptly named kingfisher hide and hobbies often hunt dragonflies and sand martins over the reserve on warm summer days. Cetti's warbler and water rail breed, roe deer are regularly seen on the open grasslands and both otters and grass snakes hunt in the deep ditches and reedbed edges.

Between late April and late August, the meadows are a sea of colour with a variety of flowering plants including several species of orchid and the very rare grass-poly, a low-growing plant that requires disturbed ground which floods in winter and dries out during the summer months.

Like all the Wildfowl and Wetland Trust centres, entry to Slimbridge is free to members and has a very well stocked shop, an excellent café and an In Focus shop that sells binoculars and telescopes. Canoe safaris run between Easter and September. The reserve is located off the A38, just north of the Gloucester and Sharpness canal near the village of Slimbridge.

Slimbridge WWT

Shapwick Heath

39. AVALON MARSHES

Lying at the heart of Somerset's Levels and Moors between the Mendip Hills to the north and the Poldens ridge in the south, the Avalon Marshes is nationally important for its rich cultural heritage as well as its diversity of wildlife. Inhabited since the end of the last ice age, it has slowly been reclaimed from the sea to create a mosaic of freshwater wetland habitats that are amongst the best in Britain.

Avalon Marshes includes a complex system of lowland wet grasslands, woodlands, rivers, streams, drainage channels and a cluster of renowned nature reserves including Huntspill River National Nature Reserve, Catcott Complex, Shapwick Moor, Shapwick Heath NNR, Westhay Moor NNR and Ham Wall NNR. These are owned and managed by a variety of organisations such as RSPB, Somerset Wildlife Trust, Natural England, Environment Agency and the Hawk and Owl Trust.

Short-eared owl

This site is best known for its internationally important assemblages of breeding and wintering birds. Bittern, marsh harrier and bearded tit all breed in good numbers and have been joined in recent years by very rare species such as great white egret, cattle egret, night heron and little bittern. Reed, sedge and Cetti's warblers are all common nesters as is the elusive water rail. Barn owl and hobbies regularly hunt for prey over the wet grasslands and open water respectively and calling cuckoos are constant companions for late spring and early summer visitors to the area.

Great white egret

Winter brings thousands of waterfowl to these wetlands, including wigeon, gadwall, teal and shoveler as well as waders such as lapwing, snipe and golden plover. The biggest winter attraction is the huge starling roost, often numbering hundreds of thousands of birds that has, in recent years, centred on the reedbeds at Ham Wall and Shapwick Heath. This spectacular murmuration inevitably attracts predators such as peregrine falcon, sparrowhawk and merlin as well as marsh harrier, barn and short-eared owls.

A number of nationally important mammal species are associated with the site. Otters are widespread, although difficult to see, as are water voles, harvest mice and water shrews. Brown hares are regularly seen at dawn and dusk in some of the fields, often with roe deer, a species that hides away in the wooded areas and tall reedbeds during broad daylight. Five species of reptiles can be seen here as can great crested newts and the introduced Iberian water frog.

Five species of river mussel are found living side by side, amongst them the rare depressed river mussel, as are an astonishing number of nationally important beetles. Prominent amongst these are the lesser silver diving beetle, greater silver diving beetle and flowering rush weevil. Other notable invertebrates are purple hairstreak butterfly, scarce chaser and hairy dragonfly, and moths such as red leopard and scarlet tiger.

Roe deer

Such an impressive array of wetland habitats inevitably support scarce plants, and these range from the rather insignificant veilwort and rootless duckweed to larger species such as milk parsley, water violet, marsh fern, marsh pea and frogbit.

Avalon Marshes is a large complex of sites lying north of the A39 between Puriton and Glastonbury. There is a visitor centre, with café, local crafts gallery, toilets and information at the Avalon Marshes Centre, Shapwick Road, just down the road from Shapwick Heath NNR although there is also a smaller one at the RSPB's Ham Wall reserve. Most of the surrounding towns and villages have a variety of cafes, pubs and restaurants.

Water vole

40. THE LIZARD PENINSULA

The most southerly point of Great Britain, the peninsula is known for its geology and its rare plants as well as its stunning scenery. Lying within the Cornwall Area of Outstanding Natural Beauty, the name is believed to be a corruption of the Cornish name 'Lys Ardh', meaning high court, and not a reference to the area's serpentine-bearing rock.

The whole peninsula covers an area of approximately 14 by 14 miles and contains several nature reserves and National Nature Reserves such as Predannack, Windmill Farm, Mullion Island, Goonhilly Downs, East Lizard Heathlands and West Lizard. Surrounded by sea, almost frost-free in winter, baked dry in the summer, it's no wonder the Lizard is home to such a unique flora. Indeed, nearly half of all our native wildflower species can be found here.

Botanically, one of the best areas is Kynance Cove with common cliff-top flowers such as sea campion and thrift at their best in May. Two rugged valleys extend inland and it is here that bloody crane's bill, green-winged orchid, hairy greenweed, thyme broomrape and wild chives grow, often alongside rarities such as land quillwort and dwarf rush. However, the site is best known for its 14 species of clover, including long-headed and twin-headed clover, two species that are found nowhere else in Britain.

Twin-headed clover

Green-winged orchid

Kynance Cove

Chough

Elsewhere on the Lizard's cliff-tops, wild asparagus, fringed rupturewort and prostrate broom can be seen, and in late summer, the whole area is transformed into a sea of purple as four species of heather, including a Lizard speciality, the Cornish heath, come into bloom. Another specialized habitat is the shallow pools found on the heathland that dry out in the summer. These provide ideal conditions for strawberry stonewort, three-lobed water crowfoot, yellow centaury and the tiny pigmy rush which occurs nowhere else in Britain.

The Lizard is also home to one of England's rarest breeding birds, the chough, which recolonized the area in 2001 after an absence of more than 30 years. Peregrine falcon, kestrel, raven and herring gull nest along the steep cliffs, as do small numbers of razorbill, guillemot and kittiwake. Dartford warblers have been breeding in suitable gorse-dominated sites since the 1980s, areas where stonechat are often common.

Its location means that the Lizard peninsula is a great place for scarce migrants in spring and autumn, and also for seawatching. Gannets and Manx shearwaters can usually be seen offshore between spring and autumn, replaced in winter by birds such as great northern and red-throated diver. Grey seals are present offshore all year round whereas spring and summer is the best time to look out for harbour porpoise and basking sharks, particularly in calm conditions off Lizard Point.

Adders, slow worms and common lizards can often be seen basking on stone walls and amongst the gorse and coastal heath in sunny weather. The mild climate and variety of habitats has resulted in a wealth of invertebrates including the marsh fritillary butterfly and the rare narrow-headed ant.

Cafes, pubs and places to stay can be found in the Lizard village along the A3083, at Cadgwith and at Mullion but the nearest town is Helston at the north western edge of the Lizard.

ACKNOWLEDGEMENTS

So many people have helped to get this book into print.
The following have all looked through and corrected various
sections, although any omissions or mistakes are my own.

Dave Allen, Tim Appleton, John Badley, Alison Barratt, Rebecca
Barrett, Ian Barthorpe, Stuart Benn, Ian Boyd, Jamie Boyle, Gareth
Brookfield, Nick Brooks, Lizzie Bruce, Paul Collin, Andrew Crory,
Pat Cullen, Ronan Dugan, Victoria Fellowes, Jack Fleming, Pete
Gordon, Barbara Hooper, Steve Hughes, Richard Miller, Pete
Moore, June Nelson, Luke Phillips, Chris Rollie, Norrie Russell,
Tony Serjeant, Martha Smith, Xanthippe Stride, Laura Thomas,
Brydon Thomason, Paul Turner, Steve Walker and Gregory
Woulahan. If I have missed anyone out, I apologise profusely.

I am very grateful to Mick Felton and the staff at Seren for
pushing me to write this tome as a follow-up to *Wild Places*,
the book on my 40 favourite Welsh sites, and for their patience
and perseverance throughout. So many photographers have
contributed stunning images to greatly improve the book and
I am indebted to them all, especially as my own pathetic
photographic efforts would never make it past the publishers.

A huge thank you has to go to my mum and dad and taid,
my wonderful grandfather, who encouraged my interest in
wildlife as a young boy. Not once did the try to steer my focus
away from the natural world, even when I was supposed to be
at school!

The biggest 'thank you', however, has to go to my wonderful wife
Ceri and our two boys, Dewi and Tomos, who have endured long
absences and supported me throughout my wildlife career, let
alone whilst writing this book. My debt to all three of you is
enormous and one I will never be able to repay. Diolch o
waelod calon xx.

PHOTOGRAPH ACKNOWLEDGEMENTS

Mike Alexander 114, 115, 116-17, 118, 124, 125, 130-31, 137; Paula J Andrews 59; Rich Andrews 129b; Christina Bollen 81; Milo Bostock 152; Ian Boyd 37r, 156, 157t, 157b, 170t; Laurie Boyle 148r; Nick Bramhall 37l; Phil Brown 84, 87; Richard Brown 129t; Graham Catley 145; Richard Carter 105tl, 105tr; Kev Chapman 53t; Steve Childs 69t; Gary Cook 60-61; Gary Cornell 88-89; Guy Curbishley 182-83; Joe M Devereux 184; James Diedrick 44t; Jimmy Edmonds 96b, 142; Peter Eeles 63r, 105b, 155, 170t, 170bl, 170br; David Evans 190; Tommy Evans 135t; Eye Candy Images 66-67; Jo Garbutt 39, 68; Stephen Gidley 90t, 90b, 101m; Laurie Goodlad 14; Nick Goodrum 33b, 179; Janet Graham 138; Michael Grant 94-95; Colin Gregory 79; Fabien Guittard 10; First-Nature 21t, 21b, 51, 70, 83l, 86l, 108r, 163l, 163r, 165, 166, 169, 171; Stephen Foster 83r; Magnus Hagdorn 174; Gail Hampshire 162; Robert Harding 50, 55; Randi Hausken 36l; Steve Herring 19, 175t; Mark Hicken 40; Chris Hill 54; Jon Hawkins 135m; David Hunter 98-99; Logan Johnson 11; Martin Kemp 47; David Kilpatrick 160-61; Joanna Kruse 172-73; Tim Lambourn 52; Brett Land 140-41; Iain Leach 20, 105ml; Tom Lee 167b; Roy Leverton 31; Loop Images Ltd 42-43, 164; John Lord 74; Malcolm Macgregor 41; Andy Mackay 144; Lyndsey Maiden 135b; Charlie Marten 28b; Vince Massimo 53b, 63l, 154t; Ian McFegan 178; Melissa McMasters 155b; Tim Melling 75; Chris Morris 85; John Morrison 44-45; Philip Mugridge 46; National Trust Photography 102-03; Natural Resources Wales 126t, 126m, 126b; Nature Photographers Ltd 187l; Andy Newman 168; Tony Parkin 23, 82; James Petts 167t; John Potter 56-57; Ben Porter 112, 113, 123t, 139; Ian Preston 143; Christopher Price 119, 120-21; RSPB 136; Alastair Rae 25; S. Rae 24t, 86r; Mike Read 30, 38; Noel Reynolds 93; Ben Sale 105mr; Iain Sarjeant 18; Rob Simmonds 149; Mark Sisson 12-13, 28c, 69b, 91, 92, 153, 158, 185; Martin Smith 58; Richard Smith 111; Jacob Spinks 177; Peter Stevens 106; Andy Sutton 8-9; Mick Sway 82; T.M.O. Landscapes 150-51; Martin Thomas Photography 34-35; Thomas Thompson 76-77; David Tipling Photo Library 26, 180-81; Chris Townsend 22; Peter Trimming 28t, 175b, 186b; Martyn Williams 146-47; Ray Wilson Bird Photography 15, 16, 71, 78, 80, 15996t, 97, 101t, 101b, 148l; Ian Woolcock 188-89; Jon Wornham 107, 108tl, 108tr, 109l, 109r, 110; Len Worthington 36r, 62r; G Xulescu 62l; Steve Young 45b

Alamy Stock Photography was the source of images on pages 8-9, 18, 26, 30, 38, 40, 42-43, 45b, 46, 50, 54, 55, 56-57, 60-61, 81, 88-89, 94-95, 98-99, 140-41, 145, 146-47, 160-61, 164, 172-73, 180-81, 182-83, 188-89.

Ian Boyd's images are to be found on www.ianboydphotography.co.uk
The First Nature website www.first-nature.com is a mine of natural images
Laurie Goodlad's image is from www.shetlandwithlaurie.com
Logan Johnson's image is from https://logansnatureblog.blogspot.com/
Tim Lambourn's image is from www.scotlandoffthebeatentrack.com
Mark Sisson's images are to be found at www.natures-images.co.uk
Chris Townsend's image is from www.christownsendoutdoors.com
Ray Wilson's images are to be found at www.raywilsonbirdphotography.co.uk
The images by Richard Carter, Peter Eeles, Iain Leech, Roy Leverton and Vince Massimo come from www.ukbutterflies.co.uk which passes on all fees to Butterfly Conservation, while the images by Stephen Foster and Andy Mackay are from www.ukmoths.org.uk